Convection-Diffusion Problems

An Introduction to Their Analysis and Numerical Solution

GRADUATE STUDIES IN MATHEMATICS 196

Convection-Diffusion Problems

An Introduction to Their Analysis and Numerical Solution

Martin Stynes
David Stynes

AMS
American Mathematical Society
Providence, RI
Atlantic Association for Research in the Mathematical Sciences
Halifax, Nova Scotia, Canada
AARMS

2010 *Mathematics Subject Classification.* Primary 65L11; Secondary 34B05, 34D15, 35B25.

For additional information and updates on this book, visit
www.ams.org/bookpages/gsm-196

Library of Congress Cataloging-in-Publication Data

Names: Stynes, M. (Martin), 1951– author. | Stynes, David, author.
Title: Convection-diffusion problems : an introduction to their analysis and numerical solution / Martin Stynes, David Stynes.
Description: Providence, Rhode Island : American Mathematical Society ; Halifax, Nova Scotia, Canada : Atlantic Association for Research in the Mathematical Sciences, [2018] | Series: Graduate studies in mathematics ; volume 196 | Includes bibliographical references.
Identifiers: LCCN 2018035503 | ISBN 9781470448684 (alk. paper)
Subjects: LCSH: Differential equations–Numerical solutions. | Diffusion–Mathematical models. | AMS: Numerical analysis – Ordinary differential equations – Singularly perturbed problems. msc | Ordinary differential equations – Boundary value problems – Linear boundary value problems. msc | Ordinary differential equations – Stability theory – Singular perturbations. msc | Partial differential equations – Qualitative properties of solutions – Singular perturbations. msc
Classification: LCC QA377 .S8785 2018 | DDC 519.2/33—dc23
LC record available at https://lccn.loc.gov/2018035503

10 9 8 7 6 5 4 3 2 1   23 22 21 20 19 18

Contents

Preface

Convection-diffusion problems attract much attention in the research literature. For numerical analysts working in this area, a standard reference is the text by Roos, Stynes, and Tobiska [**RST96**, **RST08**]. This book contains a lot of useful information, but it is daunting for those beginners who have some familiarity with numerical methods and their analysis but who have not previously worked with convection-diffusion and other singularly perturbed differential equations. For many years I felt that an easier, more introductory book was needed to encourage new people to enter our fascinating research area. This belief was encouraged by the popularity of a survey article, "Steady-state convection-diffusion problems", that I wrote for *Acta Numerica* in 2005 [**Sty05**]. The present book is an extended and updated version of that 2005 article, and I have added exercises and other material to try to make it more attractive and more useful for the novice reader.

The feeling that a book of this type was desirable did not lead me to take any action until I was invited to present a course on this topic at the AARMS (Atlantic Association for Research in the Mathematical Sciences) Summer School at Dalhousie University in Halifax, Nova Scotia, Canada, during July 2015. The organisers encourage their lecturers to transform their lecture notes into books, and after much delay I have done this. I am very grateful to AARMS for their invitation to lecture and for the enjoyable month I spent in the delightful city of Halifax.

Here we list the prerequisites for the reader. In Chapters 1–3 some knowledge of two-point boundary value problems and their numerical solution by finite difference methods is enough for almost all of the material.

For Chapter 4 it is desirable to have some previous experience of partial differential equations. Chapter 5 uses only ideas from earlier chapters. Finite element methods (FEMs) appear for the first time in the long Chapter 6, and here I assume that the reader already has a general understanding of how FEMs are constructed and analysed. The Lebesgue spaces $L^p(\Omega)$ and the standard Sobolev spaces $H^k(\Omega)$ are used occasionally in the earlier chapters of the book and more heavily in Chapter 6; the reader should have some familiarity with these well-known concepts.

The book was written where I work, in the research paradise known as Beijing Computational Science Research Center. I owe a great debt to CSRC's director Hai-Qing Lin for the positive environment he has created at CSRC through his friendly yet no-nonsense approach to productive research. My work was supported by the 1000 Talents (Foreign Experts) Program of the People's Republic of China.

All comments on this book are welcome. No doubt it will (inevitably) contain some mistakes, so corrections are also welcome, though the fewer the better! My email address is `m.stynes@csrc.ac.cn`

Martin Stynes

Chapter 1

Introduction and Preliminary Material

1.1. A simple example

Example 1.1. Consider the two-point boundary value problem

$$-\varepsilon u''(x) + 2u'(x) = 3 \quad \text{for } 0 < x < 1, \tag{1.1a}$$

$$u(0) = u(1) = 0, \tag{1.1b}$$

where ε is a small positive parameter. (This means that ε is a constant, but we are interested in what happens for different values of this constant.) The solution of (1.1) is

$$u(x) = \frac{3}{2}\left[x - \frac{e^{-2(1-x)/\varepsilon} - e^{-2/\varepsilon}}{1 - e^{-2/\varepsilon}}\right] \qquad \text{for } 0 \le x \le 1. \tag{1.2}$$

A graph of this solution for three different values of ε is displayed in Figure 1.1.

When x is not near 1, the graph is unremarkable: it appears to be the straight line $y = 3x/2$. But near $x = 1$ the solution $u(x)$ changes rapidly, and this behaviour becomes more extreme as ε gets smaller. We say that $u(x)$ has a *boundary layer* at $x = 1$ when ε is small; this is a narrow region where u is bounded independently of ε but where its derivatives are large. In fact all derivatives at $x = 1$ are unbounded as $\varepsilon \to 0$. These statements can be verified using (1.2): as $0 \le [e^{-2(1-x)/\varepsilon} - e^{-2/\varepsilon}]/[1 - e^{-2/\varepsilon}] \le 1$, we have

$$|u(x)| \le \frac{3}{2}(1+1) = 3 \qquad \text{for all } x \in [0,1],$$

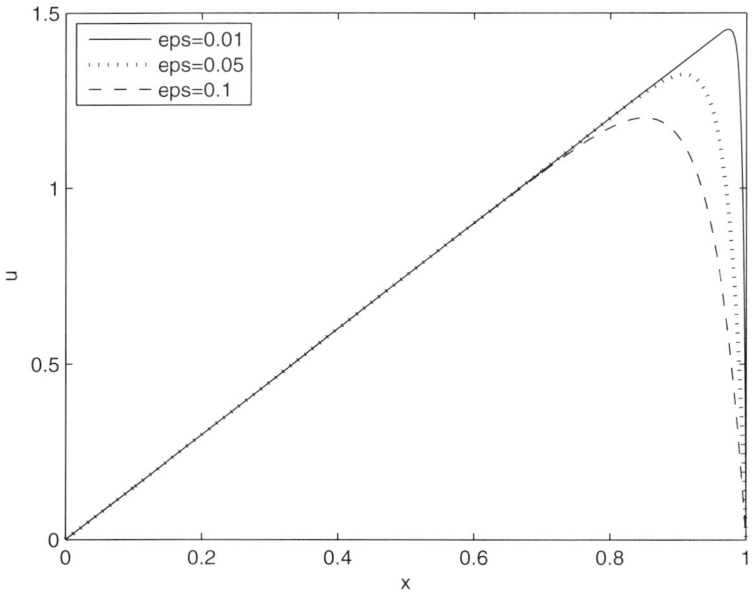

Figure 1.1. Graph of (1.2) with $\varepsilon = 0.1, 0.05, 0.01$

but

$$u'(x) = \frac{3}{2}\left[1 - \frac{2e^{-2(1-x)/\varepsilon}}{\varepsilon(1-e^{-2/\varepsilon})}\right] \approx \frac{3}{2} - \frac{3}{\varepsilon} \quad \text{for } x \text{ near } 1,$$

$$u''(x) = -\frac{6e^{-2(1-x)/\varepsilon}}{\varepsilon^2(1-e^{-1/\varepsilon})} \approx -\frac{6}{\varepsilon^2} \quad \text{for } x \text{ near } 1,$$

etc.

Exercise 1.2. Suppose that in Example 1.1 the boundary condition at $x = 1$ is changed to $u(1) = k$ for some $k \in \mathbb{R}$. Will the solution of the example still have a boundary layer at $x = 1$? Show that there is a single exceptional value of k for which the behaviour of the solution is somewhat different.

Exercise 1.3. Consider the boundary value problem

$$-\varepsilon v''(x) + v(x) = 2 \quad \text{for } 0 < x < 1,$$
$$v(0) = v(1) = 0.$$

Here, as usual, ε is a small positive parameter. Find a formula for the exact solution $v(x)$. Does v have a boundary layer when ε is small? Where?

This problem has some similarity to Example 1.1, but there are also some significant differences that will be discussed in Remark 2.37.

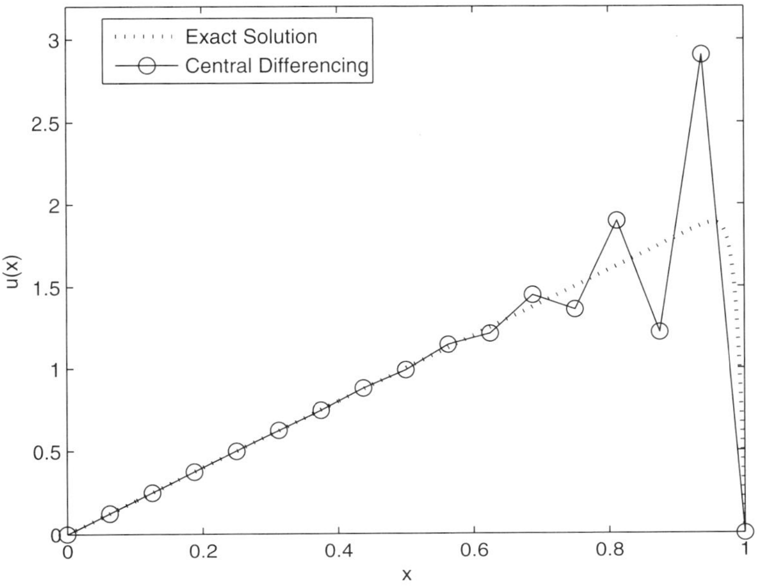

Figure 1.2. Example 1.1 with $\varepsilon = 0.01$; solution computed by central differencing with $N = 16$

The boundary value problems in this book are, like (1.1), second-order differential equations whose highest-order derivative is multiplied by a small positive parameter. Their solutions, like (1.2), are usually well-behaved on most of the domain but change rapidly near some boundary, and derivatives of the solution are large in these boundary layers.

If (1.1a) has variable coefficients or is a partial differential equation, often we are unable to write down an explicit simple formula like (1.2) for $u(x)$. Thus we need to devise analytical tools for analysing problems like (1.1) to answer questions such as:

(i) Is there a boundary layer?

(ii) At which part of the boundary is it located?

(iii) What is its analytical structure? For example, inside this layer, does $u'(x)$ behave like $1/\varepsilon$ or $1/\sqrt{\varepsilon}$ or $\cdots$?

The results of these theoretical investigations will help us devise numerical methods suitable for solving (1.1) and its generalizations, as standard numerical methods often behave poorly when boundary layers are present. To illustrate this failing, in Figure 1.2 we reproduce Figure 3.1, which displays the oscillatory and inaccurate numerical solution obtained when a routine finite difference method is used to solve Example 1.1.

1.1.1. What are convection-diffusion problems? Our interest is in elliptic operators (this term is defined in a moment) whose second-order derivatives are multiplied by some parameter ε that is allowed to be close to zero. These derivatives model diffusion, while first-order derivatives (which in this book are usually assumed to be present) are associated with convective or transport processes. In classical boundary value problems where ε is not close to zero, diffusion is the dominant mechanism in the model, and the first-order convective derivatives play a relatively minor role in the analysis. On the other hand, when ε is near zero and the elliptic differential operator has convective terms, it is called a convection-diffusion operator because now the convection term also has a strong influence on the solution of the boundary value problem. Such operators, while still satisfying the definition of ellipticity, live dangerously by flirting with the nonelliptic world. Their convective terms play a significant part in the theoretical and numerical analysis of the boundary value problem and cannot be summarily dismissed as "lower-order terms".

We shall see that the solutions of convection-diffusion problems have a convective nature on most of the domain of the problem, and the diffusive part of the differential operator is influential only in certain narrow subdomains. In these subdomains the gradient of the solution is large: its magnitude is proportional to some negative power of the parameter ε. We describe such behaviour by saying that the solution has a layer.

Definition 1.4. Suppose $v = v(x, \varepsilon)$ for $(x, \varepsilon) \in [0, 1] \times (0, 1]$. Assume that $v'(x, \varepsilon) := \partial v(x, \varepsilon)/\partial x$ exists for all $(x, \varepsilon) \in [0, 1] \times (0, 1]$. We say that v has a *layer* at a point $z \in [0, 1]$ as $\varepsilon \to 0^+$ if the following two conditions are satisfied:

(i) $\lim_{\varepsilon\to 0^+} v'(z, \varepsilon)$ is ∞ or $-\infty$;

(ii) $\lim_{\varepsilon\to 0^+} v'(x, \varepsilon)$ exists and is finite at each point $x \in [0, 1]$ satisfying $0 < |x - z| < k$ for some positive constant k, where k can depend on z but not on ε.

Remark 1.5 (Alternative definitions of a layer)**.** Definition 1.4 is, more precisely, *a layer in* v' as this is the lowest-order derivative that becomes very large as $\varepsilon \to 0$. In some situations a layer may appear only in a higher-order derivative: for instance, the Neumann boundary condition in Remark 2.33 and Exercise 2.34 induces a layer in the *second*-order derivative; see also Exercise 2.3. Such layers are less dangerous (when computing numerical solutions) than layers in first-order derivatives, but they are not entirely trouble-free. Throughout this book we shall focus our attention on the first-order derivative layers of Definition 1.4 since these are the most difficult to handle numerically.

Remark 1.6 (Structure of solutions of convection-diffusion problems). Example 1.1 is a typical example of a convection-diffusion problem. On most of the domain the solution $u(x)$ is essentially $3x/2$, which is the solution of the convective *first-order* initial value problem $2v'(x) = 3,\ v(0) = 0$. But near $x = 1$, the rapidly decaying layer function $\exp(-2(1-x)/\varepsilon)$ kicks in (the terms $\exp(-2/\varepsilon)$ in (1.2) are extremely small and can be ignored). This function is a solution of the *second-order* equation $-\varepsilon w'' + 2w' = 0$. The solution $u(x)$ has a layer at $x = 1$ in the sense of Definition 1.4.

Solutions of convection-diffusion problems throughout the book have a similar structure: on most of the domain, one sees the solution of a first-order differential equation that is obtained by setting $\varepsilon = 0$ in the differential equation, but near certain parts of the boundary this first-order solution is augmented by a layer function that is a multiple of a solution of the homogeneous second-order differential equation obtained by setting to zero the right-hand side of the original differential equation. (*Note.* In solutions, *interior layers* can occur also—that is, in Definition 1.4 one could have $z \in (0,1)$. We shall encounter this phenomenon in Remark 2.42.)

The fact that the elliptic nature of the differential operator is disguised on most of the domain means that numerical methods designed for elliptic problems will not work satisfactorily. In practice they usually exhibit a certain degree of instability. The challenge then is to modify these methods into a stable form without compromising their accuracy.

Definition 1.7. A second-order differential operator in n variables $x_1, \dots, x_n$ whose highest-order derivatives are

$$-\sum_{i,j=1}^{n} p_{ij} \frac{\partial^2(\cdot)}{\partial x_i \partial x_j},$$

where the p_{ij} are functions of $x := (x_1, \dots, x_n)$, is said to be *elliptic* on a domain (open connected set) $D \subset \mathbb{R}^n$ if

$$\sum_{i,j=1}^{n} p_{ij}(x)\xi_i\xi_j \geq \sigma \sum_{i=1}^{n} \xi_i^2 \quad \text{for all } x \in D \text{ and all } \xi_i \text{ and } \xi_j \in \mathbb{R}, \tag{1.3}$$

for some positive constant σ, which is called the ellipticity constant. (In the one-dimensional case, this definition says merely that the coefficient of u'' is negative and bounded away from zero on the domain of the problem.)

The differential operators in convection-diffusion problems stretch this definition as far as they dare: their ellipticity constant (for example, ε in Example 1.1) is close to zero.

It is often assumed (certainly in introductory textbooks in both theoretical differential equations and numerical analysis) that σ is not close to zero; for example the Laplacian $-\Delta u$ has $\sigma = 1$. This assumption avoids many difficulties. Consider, say, the proof of convergence of a finite difference method for the problem $-\sigma u''(x) + u'(x) = f(x)$ on $(0, 1)$ with $u(0) = u(1) = 0$: if you allow the positive constant σ to take a value near zero, does the argument still work? In fact, on a more fundamental level, what happens to the solution u of this boundary value problem when σ becomes small? Taking into account this alteration in the behaviour of u, how can we modify the numerical method so that it remains stable and accurate? It is questions such as these that will preoccupy us throughout this book.

Our task now is to make concrete these suspicions and assertions. We shall begin in sections 1.4 and 1.5 by developing some fundamental ideas about maximum principles and asymptotic expansions. In Chapter 2 we use these tools to begin an examination of the asymptotic nature of solutions to convection-diffusion problems. Furthermore, to carry out any numerical analysis, one needs a priori to have some bounds on the derivatives of the solutions of these problems; such estimates, and useful decompositions of the solutions, are also given in this chapter. Finite difference methods and the accuracy of their solutions are examined in Chapter 3. This leads naturally to the question of constructing suitable meshes for convection-diffusion problems, and section 3.4 is devoted to an epitome of this class: Shishkin meshes. We present in Chapter 3 a full analysis of a finite difference method on a Shishkin mesh.

The discussion up to this point has dealt only with ordinary differential equations, where the theory is fairly complete. Now we move into deeper waters: in Chapter 4 we discuss the nature of solutions to convection-diffusion problems posed in two-dimensional domains. A priori estimates for such problems are presented in section 4.2, then some preliminary comments on numerical methods are given in section 4.3. Finite difference methods for such problems are considered in Chapter 5, but our main emphasis is on Chapter 6, which is devoted to finite element methods where we shall discuss the standard Galerkin method and various stabilized finite element methods for convection-diffusion problems.

For reasons of length it is impossible to give here a complete account of the many numerical methods used to solve steady-state convection-diffusion problems. The books by Linß [**Lin10**] and Roos, Stynes, and Tobiska [**RST08**] give a comprehensive discussion of numerical methods in this active area, and Roos [**Roo12**] describes some more recent developments.

1.2. A little motivation and history

Perhaps the most common source of convection-diffusion problems is as linearizations of Navier–Stokes equations with large Reynolds number. Morton [**Mor96**] points out that this is by no means the only place where they arise: his opening chapter lists ten examples involving convection-diffusion equations that include the drift-diffusion equations of semiconductor device modelling and the Black–Scholes equation from financial modelling. He also observes that "Accurate modelling of the interaction between convective and diffusive processes is the most ubiquitous and challenging task in the numerical approximation of partial differential equations".

The numerical solution of convection-diffusion problems goes back to the 1950s [**AS55**], but only in the 1970s did it acquire a research momentum that has continued to this day. A potted history of the development of numerical methods for convection-diffusion problems up to 2003 is presented in [**Sty03b**]. The field is still very active but much remains to be done.

1.3. Notation

Throughout this book, the parameter ε lies in $(0,1]$, and we are usually interested in what happens when ε is close to zero. We shall use C to denote a generic constant that is independent of ε and of any mesh used—it can take different values in different places (even sometimes in the same calculation). A subscripted C (e.g., C_1) is also a constant that is independent of ε and of any mesh used, but it takes one fixed value.

Let g be defined on a domain $D \subset \mathbb{R}^n$ for some $n \in \{1,2,\dots\}$. In the rest of this paragraph we assume that each norm of g is well-defined and finite. We define the L_∞ norm of g by

$$\|g\|_\infty = \max_{x\in\bar{D}} |g(x)|,$$

where $\bar{D}$ denotes the closure of D. For $k = 0,1,2,\dots$ define the Sobolev seminorms and norms

$$|g|_k = \left[\int_{x\in D} \sum_{\alpha_1+\alpha_2+\cdots+\alpha_n=k} \left(\frac{\partial^k g(x)}{\partial x_1^{\alpha_1} \partial x_2^{\alpha_2} \cdots \partial x_n^{\alpha_n}} \right)^2 dx \right]^{1/2}$$

and

$$\|g\|_k = \left(\sum_{0\le j\le k} |g|_j^2 \right)^{1/2}.$$

Thus in particular $\|\cdot\|_0 \equiv \|\cdot\|_{L_2}$. Often we do not write down the domain D over which the norm is computed since typically this is obvious from the context in which the norm is used.

For any set $S \subset \mathbb{R}^n$, let ∂S denote the boundary of S. Let $C^k(S)$ denote the space of functions that are defined on S and whose derivatives up to order $k \in \{1, 2, \dots\}$ are continuous on S. Furthermore, $C(S)$ is the space of functions that are continuous on S.

1.4. Maximum principle and barrier functions

Consider the second-order differential operator L in n variables defined on some bounded domain (open connected set) $D \subset \mathbb{R}^n$ by

$$Lu(x) = -\sum_{i,j=1}^{n} p_{ij}\frac{\partial^2 u(x)}{\partial x_i \partial x_j} + \sum_{i=1}^{n} q_i(x)\frac{\partial u(x)}{\partial x_i} + r(x)u(x), \ \ x \in D,$$

where the p_{ij} are constants. Throughout section 1.4 we assume that L is elliptic in the sense of Definition 1.7.

Lemma 1.8 (Maximum Principle). *Let $u \in C(\bar{D}) \cap C^2(D)$ satisfy the differential inequality $Lu \geq 0$ on D. Suppose also that $u \geq 0$ on ∂D. Suppose that the functions q_i and r are bounded on D and that $r \geq 0$ is bounded on D. Then $u \geq 0$ on $\bar{D}$.*

This well-known result is proved, for instance, in Protter and Weinberger [**PW84**, Chapter 2, Section 3]. It is a very useful tool when analysing the behaviour of solutions to convection-diffusion problems.

Remark 1.9. In the particular case where $D = (a, b) \subset \mathbb{R}$ and $Lu = -u''$, the hypothesis $Lu \geq 0$ on D means that the graph of u is concave down on $[a, b]$. Combining this property with $u(a) \geq 0$, $u(b) \geq 0$, it is clear from a diagram (see Figure 1.3) that $u \geq 0$ on $[a, b]$.

Exercise 1.10. For the case $n = 1$ and $D = (a, b)$, let $u \in C[a, b] \cap C^2(a, b)$ with $u(a) \geq 0$, $u(b) \geq 0$. Prove Lemma 1.8 under each of the following slightly stronger hypotheses:

(1) $Lu \geq 0$ and $r > 0$ on D;

(2) $Lu > 0$ and $r \geq 0$ on D.

Hint. Assume that the conclusion of the lemma is false and derive a contradiction, using some basic calculus and the fact that a continuous function on a closed interval attains its minimum at some point in the interval.

The proof of Lemma 1.8 for the most general case $Lu \geq 0$ and $r \geq 0$ is more difficult; see [**PW84**].

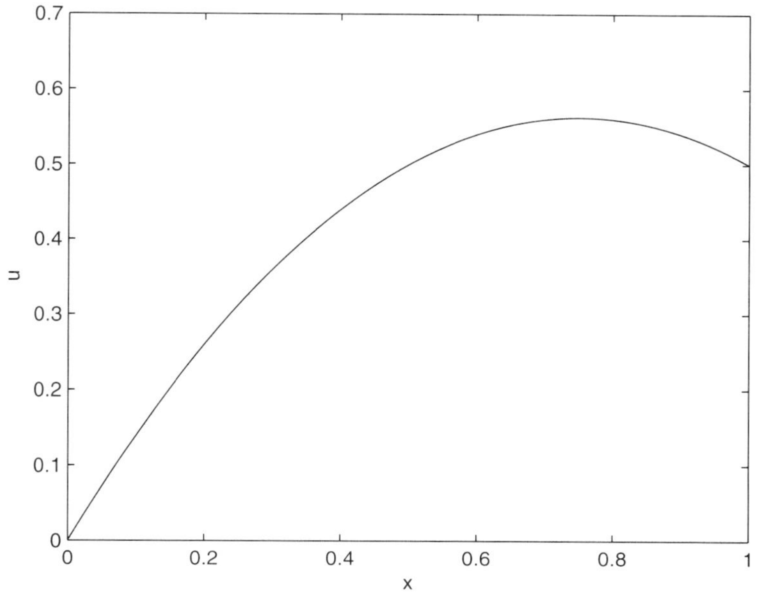

Figure 1.3. $-u'' \geq 0$ on $(0,1)$, $u(0) \geq 0, u(1) \geq 0 \implies u \geq 0$ on $[0,1]$

Exercise 1.11. For the case $n = 1$ and $D = (a,b)$, suppose that $u \in C^2[a,b]$ and that the hypothesis $u(b) \geq 0$ of Lemma 1.8 is replaced by the hypothesis $u'(b) \geq 0$. Assume that $Lu \geq 0$ and $r \geq 0$ on D, where at least one of these inequalities is strict (in fact $Lu \geq 0$ and $r \geq 0$ is sufficient, but then the proof is more complicated). Prove that the conclusion of the lemma is still valid by imitating the proof of Exercise 1.10.

Can you see what hypothesis on $u'(a)$ could replace the hypothesis $u(a) \geq 0$, while preserving the validity of the lemma? (It isn't "$u'(a) \geq 0$"!)

A maximum principle can be used to bound the absolute value of the unknown solution of a differential equation:

Corollary 1.12 (Comparison principle). *Suppose that the functions q_i and r are bounded on D and that $r(x) \geq 0$ is bounded on D. Let $u, v \in C(\bar{D}) \cap C^2(D)$. Suppose that $|Lu(x)| \leq Lv(x)$ for all $x \in D$ and $|u(x)| \leq v(x)$ for all $x \in \partial D$. Then $|u(x)| \leq v(x)$ for all $x \in \bar{D}$.*

Proof. One cannot immediately apply Lemma 1.8 to the functions $|u|$ and v because $|u|$ may not be differentiable. Instead we apply this lemma to the functions $v - u$ and $v + u$ and deduce that $v - u \geq 0$ and $v + u \geq 0$ on $\bar{D}$. But for each $x \in \bar{D}$, one has $v(x) - |u(x)| = (v - u)(x)$ or $(v + u)(x)$, so we are done. □

A function such as v in Corollary 1.12 is called a *barrier function* for u. This corollary is often applied to a function u that is a solution of a boundary value problem—so $u|_{\partial D}$ and Lu are known, but $u|_{D\setminus\partial D}$ is unknown. We then try to choose a suitable function v that satisfies the hypotheses of the corollary in order to deduce some worthwhile information about the behaviour of u inside D.

Exercise 1.13. In each of the following boundary value problems, assume that $f \in C[0,1]$, with $u(0) = u(1) = 0$. Consequently, $u \in C^2(0,1) \cap C[0,1]$. Construct an appropriate barrier function, and use a maximum principle to deduce the desired bound on $\|u\|_\infty$.

(i) If $-u'' = f$ on $(0,1)$, show that $\|u\|_\infty \le \dfrac{1}{8}\|f\|_\infty$.

(ii) If $-\varepsilon u'' + 2u' = f$ on $(0,1)$, show that $\|u\|_\infty \le \dfrac{1}{2}\|f\|_\infty$.

(iii) If $-\varepsilon u'' + 3u = f$ on $(0,1)$, show that $\|u\|_\infty \le \dfrac{1}{3}\|f\|_\infty$.

1.5. Asymptotic expansions

Putting barrier functions aside for the moment, we turn our attention to a useful descriptive tool: *asymptotic expansions*.

Let $\varepsilon > 0$ be a small parameter. If $f = f(x,\varepsilon)$ and $g = g(x,\varepsilon)$ with x lying in some domain D, we write $f(x,\varepsilon) = \mathcal{O}(g(x,\varepsilon))$ as $\varepsilon \to 0$ if there exists a positive number A that is independent of ε and an $\varepsilon_0 > 0$ such that $|f(x,\varepsilon)| \le A|g(x,\varepsilon)|$ for $0 < \varepsilon \le \varepsilon_0$. If in addition A and ε_0 are independent of x, we say that $f(x,\varepsilon) = \mathcal{O}(g(x,\varepsilon))$ as $\varepsilon \to 0$ uniformly for $x \in D$.

This notation is useful for comparing functions of similar size. For functions of greatly differing relative size, one uses a "small o" notation: we write $f(x,\varepsilon) = o(g(x,\varepsilon))$ as $\varepsilon \to 0$ if, given any $\delta > 0$, there exists an $\varepsilon_0 > 0$ such that $|f(x,\varepsilon)| \le \delta|g(x,\varepsilon)|$ for $0 < \varepsilon \le \varepsilon_0$. If in addition ε_0 is independent of x, we say that $f(x,\varepsilon) = o(g(x,\varepsilon))$ as $\varepsilon \to 0$ uniformly for $x \in D$.

These concepts can be expressed as follows:

$$f(x,\varepsilon) = \mathcal{O}(g(x,\varepsilon)) \text{ means that as } \varepsilon \to 0,\ |f(x,\varepsilon)|/|g(x,\varepsilon)| \text{ is bounded,}$$

while

$$f(x,\varepsilon) = o(g(x,\varepsilon)) \text{ means that } \lim_{\varepsilon\to 0} |f(x,\varepsilon)/g(x,\varepsilon)| = 0.$$

An *asymptotic sequence* $\{\phi_n(\varepsilon)\}$, for $n = 1, 2, \ldots$, is a sequence of functions of ε such that

$$\phi_{n+1}(\varepsilon) = o(\phi_n(\varepsilon)) \quad \text{as } \varepsilon \to 0, \text{ for each } n.$$

Examples of asymptotic sequences:

(i) $1, \varepsilon, \varepsilon^2, \varepsilon^3, \ldots,$

(ii) $\varepsilon^{1/3}, \varepsilon^{2/3}, \varepsilon, \varepsilon^{4/3}, \ldots,$

(iii) $\varepsilon|\ln\varepsilon|, (\varepsilon|\ln\varepsilon|)^2, (\varepsilon|\ln\varepsilon|)^3, \ldots.$

Asymptotic sequences are the building blocks from which one constructs asymptotic expansions.

Let $u(x,\varepsilon)$ be defined for all x in some domain D and all sufficiently small ε. Let $\{\phi_n(\varepsilon)\}$, for $n = 1, 2, \ldots$, be an asymptotic sequence. Then the series $\sum_{n=1}^{N} u_n(x)\phi_n(\varepsilon)$, where N may be finite or infinite, is said to be the *asymptotic expansion* of u with respect to $\{\phi_n\}$ as $\varepsilon \to 0$, if for each $M \in \{1, \ldots, N\}$ we have

$$u(x,\varepsilon) - \sum_{n=1}^{M} u_n(x)\phi_n(\varepsilon) = o(\phi_M) \quad \text{as } \varepsilon \to 0. \tag{1.4}$$

In this case we write $u(x,\varepsilon) \sim \sum_{n=1}^{N} u_n(x)\phi_n(\varepsilon)$. This asymptotic expansion is *uniform in* D if (1.4) holds true uniformly for $x \in D$.

There are many books on asymptotic expansions. Two good but moderately advanced ones are [**KC96**, **KC81**]. Some simpler examples of convection-diffusion problems are discussed in [**Kev00**, Section 8.3].

Note that an asymptotic expansion often consists of a finite number of terms from a *divergent* infinite series. Thus in general it is not true that taking more terms in an asymptotic expansion yields a more accurate approximation of u; instead one usually takes a small number of terms (maybe only one!) in an asymptotic expansion. The question of how many terms one should take is examined in several examples in the informative paper [**Boy99**], where much practical advice is given.

To introduce our final asymptotic concept, we take a simple example involving functions of ε that have no additional dependence on a variable x.

Example 1.14. One can easily show that one solution u_ε of the algebraic equation $u_\varepsilon^2 + \varepsilon u_\varepsilon - 1 = 0$, where ε is a small positive parameter, satisfies $u_\varepsilon = 1 + \mathcal{O}(\varepsilon)$. Thus as $\varepsilon \to 0$ this solution approaches the solution $u_0^{(1)} = 1$ of the problem $u_0^2 - 1 = 0$. Similarly, the other solution of $u_\varepsilon^2 + \varepsilon u_\varepsilon - 1 = 0$ approaches the other solution $u_0^{(2)} = -1$ of $u_0^2 - 1 = 0$. Thus as $\varepsilon \to 0$, the solutions of the original problem approach the solutions of the modified problem obtained by setting $\varepsilon = 0$.

The situation is different for the solutions $v_\varepsilon^{(1)}$ and $v_\varepsilon^{(2)}$ of the equation $\varepsilon v_\varepsilon^2 + v_\varepsilon - 1 = 0$. An application of the quadratic formula and binomial theorem shows that

$$v_\varepsilon^{(1)} = 1 - \varepsilon + 2\varepsilon^2 - 5\varepsilon^3 + \cdots, \qquad v_\varepsilon^{(2)} = -\varepsilon^{-1} - 1 + \varepsilon - 2\varepsilon^2 + \cdots.$$

Hence as $\varepsilon \to 0$, one has $v_\varepsilon^{(1)} \to 1$ (the solution of the modified problem $v_0 - 1 = 0$ obtained by setting $\varepsilon = 0$), but $v_\varepsilon^{(2)} \to -\infty$.

The first part of Example 1.14 is a *regular perturbation* problem: the behaviour of the solution when the perturbation parameter ε reaches its limit value of 0 is quite similar to the behaviour when ε is near but not equal to 0. For a regular perturbation problem, the following diagram is commutative.

$$\begin{array}{ccc} \text{Problem with } \varepsilon > 0 & \xrightarrow{\text{set } \varepsilon=0} & \text{Problem with } \varepsilon = 0 \\ \Big\downarrow \text{Solve} & & \Big\downarrow \text{Solve} \\ \text{Solution when } \varepsilon > 0 & \xrightarrow{\text{set } \varepsilon=0} & \text{Solution when } \varepsilon = 0 \end{array} \tag{1.5}$$

The second part of Example 1.14 is a *singular perturbation* problem, where reaching the limit value of the parameter causes some significant change in the solution (here $v_\varepsilon^{(2)}$ is not close to $v_0 = 1$). For a singular perturbation problem, the above diagram is not commutative, i.e., the two routes by which you can travel from its top left to its bottom right will yield different answers to "Solution when $\varepsilon = 0$".

The equation $u_\varepsilon^2 + \varepsilon u_\varepsilon - 1 = 0$ is quadratic whether or not $\varepsilon = 0$. In contrast, the equation $\varepsilon v_\varepsilon^2 + v_\varepsilon - 1 = 0$ is quadratic when $\varepsilon > 0$ but becomes linear when $\varepsilon = 0$. This fundamental change in its nature when ε reaches zero makes it unsurprising that $\varepsilon v_\varepsilon^2 + v_\varepsilon - 1 = 0$ is a singularly perturbed problem as $\varepsilon \to 0^+$ (as one can verify easily using (1.5)).

As we shall see, convection-diffusion problems form a class of singular perturbation problems.

Exercise 1.15. Show, using the commutative diagram definition (1.5), that the boundary value problem of Example 1.1 is a singularly perturbed problem. (*Hint.* Does the problem obtained by formally setting $\varepsilon = 0$ *have* a solution?)

Exercise 1.16. Show, using the commutative diagram definition (1.5), that the boundary value problem

$$\begin{aligned} -u''(x) + 2\varepsilon u'(x) &= 3 \quad \text{for } 0 < x < 1, \\ u(0) = u(1) &= 0 \end{aligned}$$

is a regular perturbation problem. Here the small parameter ε multiplies a lower-order term in the differential operator and, consequently, has much less influence on the nature of the problem than it had in Example 1.1.

Remark 1.17. The definition (1.5) is suitable for classifying problems where some first-order derivative has a layer (i.e., becomes large at some point as $\varepsilon \to 0$). But sometimes a layer appears only when higher-order derivatives are considered (see Remark 1.5), and then one needs to modify or reinterpret (1.5).

Chapter 2

Convection-Diffusion Problems in One Dimension

In this chapter we examine the asymptotic nature of solutions to convection-diffusion problems in one dimension. This will provide useful insights. The behaviour of the derivatives of these solutions, which is critical for the numerical analysis that follows later, is then discussed in detail. Finally, these two lines of attack are combined in section 2.4—decompositions of solutions.

2.1. Asymptotic analysis—an extended example

To avoid complicated algebraic details, we do not begin with the most general situation but work instead with the second-order two-point boundary value problem

$$Lu(x) := -\varepsilon u''(x) + 2u'(x) = f(x) \quad \text{for } 0 < x < 1, \tag{2.1a}$$

$$u(0) = u(1) = 0, \tag{2.1b}$$

where we recall that $\varepsilon \in (0, 1]$. Assume that $f \in C^\infty[0, 1]$. When ε is small, this is a convection-diffusion problem: the coefficient of the first-order derivative is much larger in magnitude than the coefficient of the second-order derivative. It would be more precise to write $u(x, \varepsilon)$ for the solution of (2.1), but for convenience we use $u(x)$.

For convenience we assumed in (2.1b) that the Dirichlet boundary conditions are homogeneous. Inhomogeneous boundary conditions $u(0) = u_0$ and $u(1) = u_1$ can be reduced to the homogeneous case by defining

$v(x) = u(x) - (1-x)u_0 - xu_1$ and considering $Lv(x) = f(x) + 2(u_0 - u_1)$ with $v(0) = v(1) = 0$.

If one sets $\varepsilon = 0$, then the second-order differential equation (2.1a) becomes a first-order differential equation—a significant change—so one expects that (2.1) is singularly perturbed. In the $L^\infty[0,1]$ norm, a definition of singularly perturbed is that there exists $\hat{x} \in [0,1]$ (in fact $\hat{x} = 1$ for this problem) such that

$$\lim_{\varepsilon \to 0} \lim_{x \to \hat{x}} u(x) \neq \lim_{x \to \hat{x}} \lim_{\varepsilon \to 0} u(x). \tag{2.2}$$

If (2.2) holds true, then the diagram (1.5) is not commutative (when one takes an appropriate interpretation of "set $\varepsilon = 0$" in (1.5)).

Exercise 2.1. In Example 1.1, verify that (2.2) holds true if and only if $\hat{x} = 1$.

All the important features of the general problem (2.1) are also present in (1.1).

Remark 2.2. For certain exceptional combinations of the boundary conditions and f, the problem (2.1) may be regularly—not singularly—perturbed. For example, if $f(x) \equiv 2k \in \mathbb{R}$ and the boundary conditions were changed to $u(0) = 0$, $u(1) = k$, then the solution of (2.1) becomes the well-behaved function $u(x) = kx$ and (2.2) is no longer satisfied for any $\hat{x} \in [0,1]$; i.e., (2.1) is now a regular perturbation problem.

Exercise 2.3. Suppose that in (2.1) one has $f(x) \equiv 2k \in \mathbb{R}$ and $u(1) = k - \varepsilon$. Is this a regular or a singular perturbation problem? *Hint.* Although this problem is not the same as Exercise 2.34, it comes to the same conclusion.

We begin our analysis with a preliminary result, which is of some interest in its own right.

Lemma 2.4. *Consider the problem*

$$-\varepsilon u''(x) + a(x)u'(x) = f(x) \quad \textit{for } 0 < x < 1, \tag{2.3a}$$

$$u(0) = u(1) = 0, \tag{2.3b}$$

where $a(x) \geq \underline{a}$ for all $x \in [0,1]$ and some positive constant $\underline{a}$. Then $u(x)$ does not have a boundary layer at $x = 0$ when ε is small.

Proof. By the mean value theorem, $u'(z) = 0$ for some $z \in (0,1)$. Set $A(x) = \int_0^x a(t)\,dt$ for all $x \in [0,1]$. Multiply (2.3a) by the integrating factor $\varepsilon^{-1}\exp(-A(x)/\varepsilon)$, then integrate from 0 to z. This yields

$$|u'(0)| = \left| \int_0^z \frac{1}{\varepsilon} f(x) e^{-A(x)/\varepsilon}\,dx \right| \leq \|f\|_\infty \int_0^z \frac{1}{\varepsilon} e^{-\underline{a}x/\varepsilon}\,dx \leq \frac{\|f\|_\infty}{\underline{a}}.$$

By Definition 1.4, $u(x)$ cannot have a boundary layer at $x = 0$. □

The proof of Lemma 2.4 also works under the weaker hypothesis that $\int_0^x a(t)\,dt \geq \alpha x$ for all $x \in [0,1]$.

Exercise 2.5. Modify the proof of Lemma 2.4 to bound $|u'(1))|$. Is your bound reasonably sharp? (Consider Example 1.1.)

To generate an asymptotic expansion—an infinite series—for the solution $u(x)$ of a boundary value problem such as (2.1), one begins by constructing heuristically a formal expansion. Here "formal" means that during the construction of the expansion we do not worry whether our series converges or can be differentiated term-by-term; we just steam ahead and generate a series that can in principle be computed explicitly and which we propose as an asymptotic expansion. At this stage of the analysis, *nothing has been proved.* After the series has been generated formally, we *then prove rigorously* that it really is an asymptotic expansion.

We illustrate this procedure for the problem (2.1). To begin, *assume* that

$$u(x) = \sum_{n=0}^{\infty} u_n(x)\varepsilon^n. \tag{2.4}$$

(*Note.* It is not always the case that one uses *integer* powers of ε when constructing an asymptotic expansion of the solution of a singularly perturbed differential equation, but we expect that they will work here because the derivatives of the solution of Example 1.1, which has the same differential operator as (2.1a), depend on only integer powers of ε.) We set out to find the functions $u_0, u_1, \ldots$ explicitly. Substituting (2.4) into (2.1a) yields

$$-\varepsilon \sum_{n=0}^{\infty} u_n''(x)\varepsilon^n + 2\sum_{n=0}^{\infty} u_n'(x)\varepsilon^n = f(x).$$

Comparing coefficients of powers of ε, one gets

$$2u_0'(x) = f(x), \quad 2u_1'(x) = u_0''(x), \quad 2u_2'(x) = u_1''(x), \quad \text{etc.} \tag{2.5}$$

To solve in turn each of these first-order ordinary differential equations for $u_0, u_1, \ldots$, each equation should have associated with it a single boundary condition. But the boundary conditions (2.1b) seem to imply that $u_n(0) = u_n(1) = 0$ for all n: twice as many conditions as we can handle!

Expansions like (2.4) are also used in regular perturbation problems. They contain no special feature designed to handle boundary layers. Consequently, (2.4) is unable to accommodate the boundary condition at a layer, which, like Example 1.1, will be handled by a layer-type function. Thus when constructing the asymptotic expansion (2.4), *one must discard boundary conditions where a layer occurs.* Now Lemma 2.4 tells us that there is no layer at $x = 0$; thus the layer (if any) is at $x = 1$, so we should ignore the boundary condition at $x = 1$ and ask of (2.4) only that it satisfy the condition $u(0) = 0$ from (2.1b). That is, we require that $u_n(0) = 0$ for all n.

We can now solve the equations (2.5) for the $u_n(x)$:

$$u_0(x) = \frac{1}{2}\int_0^x f(t)\,dt, \quad u_1(x) = \frac{f(x) - f(0)}{2}, \quad u_2(x) = \frac{f'(x) - f'(0)}{2}, \quad \text{etc.}$$

Thus (2.4) becomes

$$\sum_{n=0}^{\infty} \left[F^{(n)}(x) - F^{(n)}(0)\right] \varepsilon^n, \quad \text{where } F(x) := \frac{1}{2}\int_0^x f(t)\,dt. \tag{2.6}$$

One can show that $u(x) = \sum_{n=0}^{M} \left[F^{(n)}(x) - F^{(n)}(0)\right] \varepsilon^n + o(\varepsilon^M)$ for each $M \geq 0$, so (1.4) is satisfied. But one finds that this asymptotic expansion is not uniform for $0 \leq x \leq 1$; it is uniform only for $0 \leq x \leq \delta$ where δ is any fixed constant in $(0, 1)$. This situation is unsatisfactory since at $x = 1$ we expect that $u(x)$ has a boundary layer, which is its most interesting feature. Of course the inadequacy of (2.4) near $x = 1$ is unsurprising because our construction of it has ignored completely the boundary condition $u(1) = 0$ from (2.1b).

If (2.1) were a regular perturbation problem, then (2.6) would turn out to be an asymptotic expansion of $u(x)$ uniformly for $0 \leq x \leq 1$. The asymptotic expansion (2.6) fails to be uniform for $0 \leq x \leq 1$ precisely because (2.1) is singularly perturbed.

Our assumption that (2.4) is sufficiently complicated to deal fully with the boundary value problem (2.1) has turned out to be false. What can be done to improve the asymptotic expansion (2.6)? Consider the special case $f(x) \equiv 3$. Then (2.6) collapses to the function $3x/2$, but the exact solution is given by (1.2). In this formula the terms $e^{-1/\varepsilon}$ are "exponentially small" (i.e., negligible compared with any integer power of ε) and can safely be ignored. What is missing from (2.6) is some approximation of $e^{-(1-x)/\varepsilon}$; that is, some function of $(1 - x)/\varepsilon$ must be added to (2.6).

A standard systematic way of introducing a function to handle a boundary layer is as follows: define the *stretched variable* $\rho := (1-x)/\varepsilon$ and rewrite the differential equation as a function of ρ instead of a function of x.

Remark 2.6. In the formula for ρ, the number 1 appears as the location of the layer, but the division by ε is more subtle. The purpose of stretching the variable is to achieve the same dependence on ε in the most significant terms of the transformed differential operator (i.e., to balance diffusion and convection inside the layer as both processes play a role there), but the exact scaling to use in general singular perturbation problems is not always obvious. Here we scale by ε because the derivatives that appear in specific examples, such as Example 1.1, involve integer powers of ε. See [**KC96**, Chapter 4] for examples with different scalings.

Thus set $\tilde{u}(\rho) \equiv u(x)$ for $0 < \rho < 1/\varepsilon$ (corresponding to $0 < x < 1$). In fact one works instead with $0 < \rho < \infty$ as the details are then slightly simpler. Now

$$\frac{du}{dx} = \frac{d\tilde{u}}{d\rho}.\frac{d\rho}{dx} = -\frac{1}{\varepsilon}\tilde{u}_\rho \quad \text{and} \quad u''(x) = \frac{1}{\varepsilon^2}\tilde{u}_{\rho\rho}\,,$$

so writing the differential operator in terms of ρ, we get

$$-\varepsilon u'' + u' = -\frac{1}{\varepsilon}\left(\tilde{u}_{\rho\rho} + \tilde{u}_\rho\right) =: \tilde{L}\tilde{u}.$$

Note how the diffusion and convection terms now have the same dependence on ε.

By its construction, the original asymptotic expansion $\sum_{n=0}^{\infty} u_n(x)\varepsilon^n$ in (2.4) satisfied $L\left(\sum_{n=0}^{\infty} u_n(x)\varepsilon^n\right) = f$, so the correction $v(\rho)$ that is to be added to this expansion must satisfy $\tilde{L}v = 0$, i.e., $v_{\rho\rho} + v_\rho = 0$. This second-order differential equation needs boundary conditions on $v(\rho)$ at both $\rho = 0$ (which corresponds to $x = 1$) and at $\rho = \infty$. We can now finally enforce the original boundary condition $u(1) = 0$ by requiring that our modified asymptotic expansion satisfies this condition, i.e., that

$$\sum_{n=0}^{\infty} u_n(1)\varepsilon^n + v(0) = 0.$$

One wants the function v to act like a boundary layer, which implies that it dies off rapidly as ρ becomes large. Thus it is natural to impose also the boundary condition $v(\infty) = 0$.

The two-point boundary value problem that defines v is now completely specified:

$$v_{\rho\rho} + v_\rho = 0 \ \text{ for } 0 < \rho < \infty, \quad v(0) = -\sum_{n=0}^{\infty} u_n(1)\varepsilon^n, \quad v(\infty) = 0.$$

This can be solved explicitly:

$$\begin{aligned} v(\rho) &= e^{-\rho}v(0) \\ &= -e^{-(1-x)/\varepsilon}\sum_{n=0}^{\infty} u_n(1)\varepsilon^n \\ &= -e^{-(1-x)/\varepsilon}\sum_{n=0}^{\infty}\left[F^{(n)}(1) - F^{(n)}(0)\right]\varepsilon^n. \end{aligned}$$

Adding this term to (2.6), the new proposed expansion is

$$u_{as}(x) := \sum_{n=0}^{\infty} \left[F^{(n)}(x) - F^{(n)}(0)\right] \varepsilon^n - e^{-(1-x)/\varepsilon} \sum_{n=0}^{\infty} \left[F^{(n)}(1) - F^{(n)}(0)\right] \varepsilon^n. \tag{2.7}$$

Up to this moment, the calculation is merely formal; nothing has been proved. Now we justify (2.7) rigorously.

To show that (2.7) is indeed a valid asymptotic expansion, i.e., that $u(x) \sim u_{as}(x)$, set

$$\theta_M(x) = u(x) - \sum_{n=0}^{M} \left[F^{(n)}(x) - F^{(n)}(0)\right] \varepsilon^n + e^{-(1-x)/\varepsilon} \sum_{n=0}^{M} \left[F^{(n)}(1) - F^{(n)}(0)\right] \varepsilon^n \quad \text{for } M = 0, 1, 2, \ldots.$$

We shall bound θ_M by means of a suitably chosen barrier function. Now $\theta_M(1) = 0$ and $\theta_M(0) = e^{-1/\varepsilon} \sum_{n=0}^{M} \left[F^{(n)}(1) - F^{(n)}(0)\right] \varepsilon^n = \mathcal{O}\left(\varepsilon^{M+1}\right)$ because of the exponentially small factor $e^{-1/\varepsilon}$. Also,

$$\begin{aligned} L\theta_M(x) &= f(x) - \sum_{n=0}^{M} \left[-\varepsilon F^{(n+2)}(x) + F^{(n+1)}(x)\right] \varepsilon^n \\ &= f(x) - F'(x) + \varepsilon^{M+1} F^{(M+2)}(x) \\ &= \varepsilon^{M+1} F^{(M+2)}(x), \end{aligned}$$

where the series telescoped. This spectacular cancellation of almost all terms always happens if the proposed asymptotic expansion has been constructed correctly. Define the barrier function $b(x) = C\varepsilon^{M+1}(1 + x)$, where the constant $C \geq \|F^{(M+2)}\|_\infty$ is chosen such that $b(0) = C\varepsilon^{M+1} \geq |\theta_M(0)|$. Trivially, $b(1) \geq |\theta_M(1)| = 0$. Finally, $Lb(x) = C\varepsilon^{M+1} \geq |L\theta_M(x)|$ for $0 < x < 1$. By Corollary 1.12, $|\theta_M(x)| \leq b(x) \leq 2C\varepsilon^{M+1}$ for $0 \leq x \leq 1$, and this is $o(\varepsilon^M)$ uniformly for $x \in [0, 1]$. Thus (2.7) is an asymptotic expansion of $u(x)$, the solution of (2.1), that is valid uniformly for $0 \leq x \leq 1$.

For more general convection-diffusion problems on $[0, 1]$, where the coefficient of u' is positive, an analysis similar to the above (see [**RST08**, pp. 12–15] for the details) will construct functions $u_n(x)$ and $v_n(x)$ such that for $k = 0, 1, 2, \ldots$, one has

$$u(x) = \sum_{n=0}^{k} u_n(x)\varepsilon^n + \sum_{n=0}^{k} v_n(x)\varepsilon^n + \varepsilon^{k+1} R(x, \varepsilon, k), \tag{2.8}$$

where $|u_n^{(i)}(x)| \le C$ and $|v_n^{(i)}(x)| \le C\varepsilon^{-i}e^{-\alpha(1-x)/\varepsilon}$ for all i and n, with $C = C(i,n)$, and $|R(x,\varepsilon,k)| \le C = C(k)$ uniformly for $0 \le x \le 1$. Hence $\sum_{n=0}^{\infty} u_n(x)\varepsilon^n + \sum_{n=0}^{\infty} v_n(x)\varepsilon^n$ is an asymptotic expansion of $u(x)$ that is valid uniformly for $0 \le x \le 1$.

Exercise 2.7. Construct an asymptotic expansion for the regular perturbation problem of Exercise 1.16, and prove that this expansion is valid uniformly for $0 \le x \le 1$.

Exercise 2.8. A singularly perturbed boundary value problem can have two boundary layers. Compute the exact solution of the *reaction-diffusion problem* (zero-order terms can model reactions in chemical processes)

$$-\varepsilon u''(x) + 4u(x) = 6 \text{ on } (0,1), \quad u(0) = u(1) = 0.$$

Observe that $u(x)$ has boundary layers at $x = 0$ *and* $x = 1$. Construct an asymptotic expansion for this problem using an appropriate asymptotic sequence (derivatives of the exact solution will guide you in what powers of ε to use—the expansion is different from (2.4)) and prove that this expansion is valid uniformly for $0 \le x \le 1$.

The construction of asymptotic expansions can be much more complicated than our extended example reveals: the exact scaling to use when stretching variables is not always easy to find, and it may take some work to determine the boundary conditions that must be satisfied by each term in the expansion. Further examples of asymptotic expansions of solutions of singularly perturbed problems can be found in [**KC96**, **O'M14**, **Smi85**, **VeBK95**]. For a comprehensive discussion of the construction of asymptotic expansions for a large variety of convection-diffusion problems in n dimensions, see [**Il'92**].

2.2. Green's functions

We assume the reader is familiar with the general usage of Green's functions as they are a standard topic in the theory of ordinary differential equations. In this section we describe aspects of their behaviour that are peculiar to convection-diffusion problems.

Consider the problem

$$Lu(x) := -\varepsilon u''(x) + a(x)u'(x) + b(x)u(x) = f(x) \quad \text{for } 0 < x < 1, \tag{2.9a}$$

$$u(0) = 0, \ u(1) = 0, \tag{2.9b}$$

where $a(x) \ge \underline{a} > 0$ for some constant $\underline{a}$, and $b(x) \ge 0$ on $[0,1]$. Assume that $a, b, f \in C[0,1]$. It then follows from the standard theory of ordinary differential equations that (2.9) has a unique solution $u \in C^2[0,1]$, and the Green's function for (2.9) exists and is unique.

For each point $x \in (0,1)$, the Green's function associated with the operator L and the point x satisfies

$$L_\xi^* G(\xi,x) = \delta(\xi - x) \quad \text{for } 0 < \xi < 1, \quad G(0,x) = G(1,x) = 0, \tag{2.10}$$

where $\delta(\cdot)$ is the Dirac δ-distribution and the adjoint L_ξ^* of L is defined by

$$L_\xi^* G(\xi,x) := -\varepsilon G_{\xi\xi}(\xi,x) - \big(a(\xi)G(\xi,x)\big)_\xi + b(\xi)G(\xi,x).$$

Here we regard G as a function of ξ with x fixed.

Then one has the key property of Green's functions:

$$u(x) = \int_0^1 G(\xi,x) f(\xi)\, d\xi.$$

In classical terms, $G(\cdot,x) \in C[0,1]$ is defined by

$$\begin{cases} L_\xi^* G(\xi,x) = 0 \quad \text{for } 0 < \xi < x \text{ and } x < \xi < 1, \\ \lim_{\xi \to x^-} G(\xi,x) - \lim_{\xi \to x^+} G(\xi,x) = 0, \\ \lim_{\xi \to x^-} G_\xi(\xi,x) - \lim_{\xi \to x^+} G_\xi(\xi,x) = 1/\varepsilon, \\ G(0,x) = G(1,x) = 0. \end{cases} \tag{2.11}$$

In particular, if $a(\cdot)$ is a positive constant and $b \equiv 0$, then (2.11) yields

$$G(\xi,x) = \begin{cases} \dfrac{[1 - e^{-a\xi/\varepsilon}][1 - e^{-a(1-x)/\varepsilon}]}{a(1 - e^{-a/\varepsilon})} & \text{for } 0 \le \xi \le x, \\ \dfrac{[e^{-a(\xi - x)/\varepsilon} - e^{-a(1-x)/\varepsilon}][1 - e^{-ax/\varepsilon}]}{a(1 - e^{-a/\varepsilon})} & \text{for } x < \xi \le 1. \end{cases} \tag{2.12}$$

Exercise 2.9. Let $a(\cdot)$ be a positive constant, and let $b \equiv 0$. Verify that the function defined in (2.12) satisfies (2.11). Conversely (which is a little more difficult), start from (2.11) and derive (2.12).

A graph of this Green's function $G(\cdot,x)$ for $0 \le \xi \le 1$ when $a \equiv 2, b \equiv 0$ and $x = 0.4$ is displayed in Figure 2.1. It has layers at the *left-hand* ends of the intervals $[0,x]$ and $[x,1]$ because the coefficient $-a(\xi)$ of the convective derivative G_ξ in (2.10) is negative; see Exercise 2.10.

Exercise 2.10. Suppose that in (2.9) one has $a(x) < 0$ on $[0,1]$. Show that the change of variable $x \mapsto 1 - x$ (which swaps the endpoints 0 and 1 of the interval $[0,1]$) will transform (2.9) to our standard boundary value problem for which $a(x) > 0$. Thus the essential nature of $u(x)$ remains unaltered when $a(x) < 0$ on $[0,1]$, except that the boundary layer is then at the left-hand boundary $x = 0$.

As is well known, an equivalent alternative definition of Green's function is to treat it as a function of x for fixed $\xi \in (0,1)$. For our problem, this

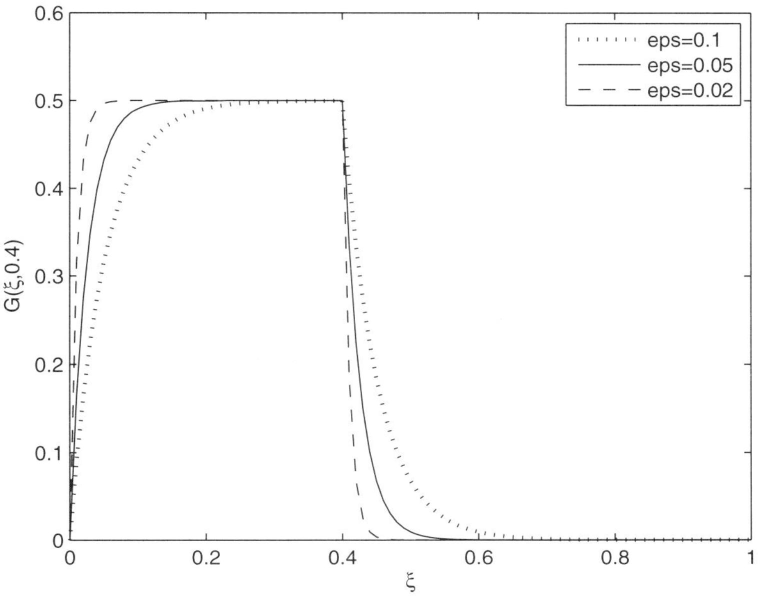

Figure 2.1. Graph of Green's function (2.12) with $a = 2$, $x = 0.4$, and $\varepsilon = 0.1, 0.05, 0.02$

definition is the following:

$$\begin{cases} LG(\xi, x) = 0 \quad \text{for } 0 < x < \xi \text{ and } \xi < x < 1, \\ \lim_{x\to\xi^-} G(\xi, x) - \lim_{x\to\xi^+} G(\xi, x) = 0, \\ \lim_{x\to\xi^-} G_x(\xi, x) - \lim_{x\to\xi^+} G_x(\xi, x) = 1/\varepsilon, \\ G(\xi, 0) = G(\xi, 1) = 0. \end{cases} \tag{2.13}$$

A graph of $G(\xi, \cdot)$, for $0 \le x \le 1$ when $a \equiv 2$, $b \equiv 0$, and $\xi = 0.3$, is given in Figure 2.2.

Exercise 2.11. Fix $\xi \in (0, 1)$. Show by a contradiction argument that if $b > 0$ and $\phi \in C[0, 1]$ satisfies

$$\begin{cases} L\phi(x) \ge 0 \quad \text{for } 0 < x < \xi \text{ and } \xi < x < 1, \\ \lim_{x\to\xi^-} \phi'(x) - \lim_{x\to\xi^+} \phi'(x) \ge 0, \\ \phi(0) \ge 0, \ \phi(1) \ge 0, \end{cases}$$

then $\phi \ge 0$ on $[0, 1]$. This is an extension of our old maximum principle, Lemma 1.8. (It remains true when one has only $b \ge 0$, but then the proof is more complicated.) Use this maximum principle twice with suitable barrier functions (you can assume the principle for $b \ge 0$) to show that for fixed ξ one has

$$0 \le G(\xi, \cdot) \le \frac{1}{a} \text{ on } [0, 1].$$

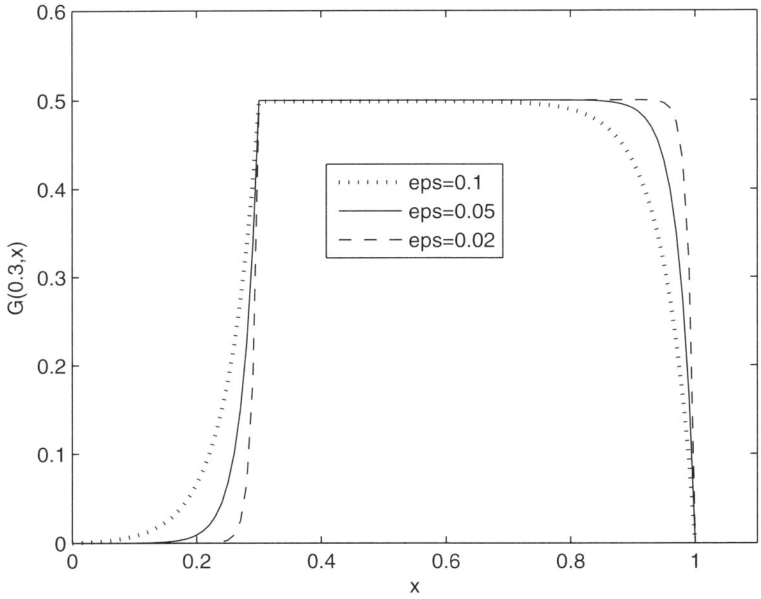

Figure 2.2. Graph of Green's function (2.12) with $a = 2$, $\xi = 0.3$, and $\varepsilon = 0.1, 0.05, 0.02$

Hint. For the upper bound on G you need to construct a barrier function $\psi(x)$ resembling Figure 2.2, but you can simplify matters by taking ψ constant for $\xi \le x \le 1$.

See [**Lin10, RST08**] for more extensive discussions of the use of Green's functions in the analysis of singularly perturbed differential equations.

2.3. A priori bounds on the solution and its derivatives

Asymptotic expansions of the solution u of a convection-diffusion problem such as (2.1) give us a good idea of how u behaves. In addition, information about the derivatives of u is needed to analyse numerical methods, so in this section we derive pointwise bounds on these derivatives. Pointwise bounds on derivatives imply bounds in other norms such as the Sobolev $H^1[0,1]$ norm.

Consider the general convection-diffusion problem

$$Lu(x) := -\varepsilon u''(x) + a(x)u'(x) + b(x)u(x) = f(x) \quad \text{for } 0 < x < 1, \tag{2.14a}$$

$$u(0) = g_0,\ u(1) = g_1, \tag{2.14b}$$

where $a(x) \ge \underline{a} > \alpha > 0$ and $b(x) \ge 0$ on [0,1], g_0 and g_1 are given constants, and one can choose values for the constants $\underline{a}, \alpha$ using the definition of $a(\cdot)$.

Assume for the moment that $a, b, f \in C[0,1]$, though sometimes more regularity will be needed in the analysis that follows. It then follows from the standard theory of ordinary differential equations that (2.14) has a unique solution $u \in C^2[0,1]$.

Remark 2.12. In fact, when $a(x) \geq \underline{a} > 0$ on $[0,1]$, then one does not need to assume also that $b \geq 0$ because this latter property can be induced by a change of variable, provided that ε is sufficiently small. To see this, set $u(x) = v(x)e^{kx}$ where the constant k is yet to be chosen. Then $Lu = f$ is equivalent to

$$-\varepsilon v''(x) + [a(x) - 2\varepsilon k]v'(x) + [b(x) + ka(x) - \varepsilon k^2]v(x) = f(x)e^{-kx}, \tag{2.15}$$

and—for $a(x) \geq \underline{a} > 0$ and ε sufficiently small—one can choose k with $0 \leq k \leq C$ (some constant C that is independent of ε) such that the coefficients of v' and v in (2.15) are both positive, so v satisfies a differential equation of the desired type. Now one can use barrier functions and Corollary 1.12 to analyse the solution v of (2.15) and, hence, obtain information about u from $u(x) = v(x)e^{kx}$.

Lemma 2.25 will give an alternative manifestation of this manoeuvre.

We shall frequently prove results under the hypothesis that "ε is sufficiently small". This assumption is not restrictive because if ε is bounded away from zero, then the problem is no longer singularly perturbed and tehniques appropriate to the case $\varepsilon = 1$ can be used.

Exercise 2.13. In Remark 2.12, verify that the equation $Lu = f$ is equivalent to (2.15) and that when ε is sufficiently small, one can choose k to satisfy the conditions stated in the remark.

Lemma 2.14. *Let u be the solution of (2.14). Set*

$$C_1 = \frac{\alpha|g_1 - g_0| + \|f\|_\infty + |g_0|\,\|b\|_\infty}{\alpha}.$$

Then

$$\|u\|_\infty \leq C_1 + |g_0| \tag{2.16}$$

and

$$|u'(0)| \leq C_1. \tag{2.17}$$

Proof. Set $z(x) = u(x) - g_0$ for $0 \leq x \leq 1$. Then $z(0) = 0$, $|z(1)| = |g_1 - g_0|$, and

$$|Lz(x)| = |f(x) - g_0 b(x)| \leq \|f\|_\infty + |g_0|\,\|b\|_\infty.$$

Apply Corollary 1.12 to bound $|z(x)|$ by the barrier function

$$\theta(x) = \frac{x}{\alpha}\big(\alpha|g_1 - g_0| + \|f\|_\infty + |g_0|\,\|b\|_\infty\big).$$

This immediately implies (2.16), and (2.17) follows from

$$|u'(0)| = \lim_{x\to 0^+} [|z(x)|/x] \le \lim_{x\to 0^+} [\theta(x)/x].$$

(Note that our barrier function θ needs to satisfy $\theta(0) = 0$ in order to bound $|u'(0)|$.) □

Inequality (2.17) shows that the solution $u(x)$ of (2.14) has no boundary layer at $x = 0$ as $\varepsilon \to 0^+$ (cf. Lemma 2.4). It will in general have a boundary layer at $x = 1$, like Example 1.1.

Exercise 2.15. Try to imitate the proof of Lemma 2.14 to bound $|u'(1)|$. You will be unable to prove $|u'(1)| \le C$, but it is instructive to see what goes wrong. From Example 1.1 one suspects that the correct bound is $|u'(1)| \le C\varepsilon^{-1}$; prove this bound by using a maximum principle with the barrier function

$$\theta_1(x) := k_1 \left[e^{-k_2(1-x)/\varepsilon} - e^{-k_3(1-x)/\varepsilon} \right],$$

where the constants k_1, k_2, k_3 have to be chosen appropriately. Show further that under the stronger hypothesis that $b \ge \beta > 0$ for some constant β, one can obtain the desired bound on $|u'(1)|$ by using the simpler barrier function

$$\theta_2(x) := k_4 \left[1 - e^{-k_5(1-x)/\varepsilon} \right]$$

for suitable constants k_4, k_5.

Away from $x = 1$, we know from section 2.1 that the solution of (2.14) satisfies $u(x) \approx u_0(x)$, where $u_0(x)$ is the solution of the *reduced problem*

$$a(x)u_0'(x) + b(x)u_0(x) = f(x) \quad \text{for } 0 < x < 1,\ u(0) = g_0. \tag{2.18}$$

This is the same $u_0(x)$ as the first term in (2.4); it is also the term $3x/2$ in (1.2). We call u_0 the *reduced solution* of the original problem (2.14).

Exercise 2.16. Use a maximum principle and barrier function argument on the interval $[0, 1]$ to show that there exists a constant C such that

$$|u(x) - u_0(x)| \le C\varepsilon \quad \text{for } 0 \le x \le 1 - (\varepsilon/\alpha)\ln(1/\varepsilon).$$

When deriving pointwise bounds on the derivatives of u, the key step is obtaining a sharp bound on $u'(x)$, so we shall spend a lot of time examining how one proves this.

We have $\|a\|_\infty \le C$ and $\|b\|_\infty \le C$. To push through an inductive proof bounding $u'', u^{(3)}, \ldots$ in Theorem 2.27 and Exercise 2.28, it emerges that when bounding $u'(x)$ from (2.14) one should replace the right-hand side $f(x)$ by the more general hypothesis that $f = f(x, \varepsilon)$ with

$$\left| \frac{\partial^i f(x,\varepsilon)}{\partial x^i} \right| \le C_0 \left(1 + \varepsilon^{-1-i} e^{-\alpha(1-x)/\varepsilon} \right) \tag{2.19}$$

for $i = 0, 1, 2, \ldots, q$ and $x \in [0,1]$, where q is some positive integer and C_0 is some fixed constant.

Lemma 2.17. *Assume* (2.19). *Then there exists a constant* C_2 *such that* $\|u\|_\infty \le C_2$ *and* $|u'(0)| \le C_2$.

Exercise 2.18. Prove Lemma 2.17 by imitating the proof of Lemma 2.14 but using the barrier function $B(x) = C_3\left[x + e^{-\alpha(1-x)/\varepsilon} - e^{-\alpha/\varepsilon}\right]$, for some constant C_3, to bound $z(x) := u(x) - g_0$. (Observe that you need $\alpha < \underline{a}$ to make the proof work.) Why can't we invoke Lemma 2.14 directly to get the desired bounds?

We begin with a nonlocalized bound on $\|u'\|_\infty$.

Lemma 2.19. *There exists a constant* C_4 *such that* $\|u'\|_\infty \le C_4\varepsilon^{-1}$.

Proof. As $u \in C^2[0,1]$, we can choose $x \in [0,1]$ such that $|u'(x)| = \|u'\|_\infty$. Assume without loss of generality that $\varepsilon \le 2\|a\|_\infty$. Choose an interval $[x_1, x_2] \subset [0,1]$ such that $x \in [x_1, x_2]$ and $x_2 - x_1 = \varepsilon/(2\|a\|_\infty)$. By the mean value theorem and Lemma 2.17, there exists $\tilde{x} \in (x_1, x_2)$ such that

$$|u'(\tilde{x})| = \left|\frac{u(x_2) - u(x_1)}{x_2 - x_1}\right| \le 4C_2\|a\|_\infty\varepsilon^{-1}.$$

Integrating (2.14) from x to $\tilde{x}$ and rearranging gives

$$\|u'\|_\infty = |u'(x)| \le |u'(\tilde{x})| + \varepsilon^{-1}\left|\int_x^{\tilde{x}} \left[|a(s)u'(s)| + |f(s)| + |b(s)u(s)|\right] ds\right|.$$

Hence, invoking (2.19) to bound f and Lemma 2.17 to bound u, and observing that $|x - \tilde{x}| \le \varepsilon/(2\|a\|_\infty)$, we get

$$\|u'\|_\infty \le 4C_2\|a\|_\infty\varepsilon^{-1} + \|u'\|_\infty/2 + C \le C\varepsilon^{-1} + \|u'\|_\infty/2.$$

The result follows by solving for $\|u'\|_\infty$. □

Lemma 2.19 gives a sharp bound on $\|u'\|_\infty$, but it does not reveal that $|u'(x)|$ is large only near $x = 1$. The proof of this layer property of u' is the main aim in the rest of section 2.3.

Theorem 2.20. *There exists a constant* C *such that*

$$|u'(x)| \le C\left[1 + \varepsilon^{-1}e^{-\alpha(1-x)/\varepsilon}\right] \quad \textit{for } 0 \le x \le 1. \tag{2.20}$$

Proof. A different proof of (2.20) is given each of the following four subsections. □

We provide four different proofs because they introduce a variety of techniques that are often useful when manipulating solutions of convection-diffusion problems. Some of the proofs demand more regularity of the data

a, b and f, so we shall where necessary make additional assumptions on this data.

2.3.1. Kellogg and Tsan technique. In [**KT78**] an integrating factor and some elementary manipulations are used to handle (2.14), as we now describe.

1st proof of Theorem 2.20. Set $h = f - bu$ and

$$A(x) = \int_0^x a(t)\,dt \quad \text{for } 0 \le x \le 1.$$

Rewrite (2.14) as $-\varepsilon u'' + au' = h$. Multiply this by the integrating factor $\varepsilon^{-1}e^{-A(x)/\varepsilon}$, then integrate from x to 1. Rearranging, we get

$$u'(x) = e^{-[A(1)-A(x)]/\varepsilon}u'(1) + \varepsilon^{-1}\int_{t=x}^1 e^{-[A(t)-A(x)]/\varepsilon}h(t)\,dt.$$

Invoking Lemma 2.19 to bound $u'(1)$, and noting that $A(s)-A(x) \ge \underline{a}(s-x)$ for $s \ge x$, it follows that

$$|u'(x)| \le C\varepsilon^{-1}e^{-\underline{a}(1-x)/\varepsilon} + \varepsilon^{-1}\int_{t=x}^1 e^{-\underline{a}(t-x)/\varepsilon}|h(t)|\,dt. \tag{2.21}$$

By (2.19) and Lemma 2.17,

$$\begin{aligned}
\varepsilon^{-1}&\int_{t=x}^1 e^{-\underline{a}(t-x)/\varepsilon}|h(t)|\,dt \\
&\le C\varepsilon^{-1}\int_{t=x}^1 e^{-\underline{a}(t-x)/\varepsilon}\left[1+\varepsilon^{-1}e^{-\alpha(1-t)/\varepsilon}\right]dt \\
&= C\left[1-e^{-\underline{a}(1-x)/\varepsilon}\right] + C\varepsilon^{-2}e^{-\alpha(1-x)/\varepsilon}\int_{t=x}^1 e^{-(\underline{a}-\alpha)(t-x)/\varepsilon}\,dt \\
&\le C\left[1+\varepsilon^{-1}e^{-\alpha(1-x)/\varepsilon}\right].
\end{aligned}$$

Recalling (2.21), we are done. □

Remark 2.21. While the above proof is short and requires only that a, b and f lie in $C[0,1]$, it does not seem possible to generalize it to problems in higher dimensions such as

(2.22a)
$$-\varepsilon\Delta u + a_1(x,y)u_x + a_2(x,y)u_y + b(x,y)u = f(x,y) \text{ on } \Omega = (0,1)^2,$$
(2.22b)
$$u = 0 \text{ on } \partial\Omega,$$

where $a_1 > 0,\ a_2 > 0$, and $b \ge 0$ on $\bar{\Omega}$.

Exercise 2.22. Suppose that (2.19) is replaced by the stronger inequality $|f(x,\varepsilon)| \le C_0$, so f is better behaved. Show that (2.20) can then be improved to $|u'(x)| \le C[1+\varepsilon^{-1}e^{-\underline{a}(1-x)/\varepsilon}]$ for $0 \le x \le 1$.

2.3.2. Majorizing function (barrier function) approach. This elementary method generalizes the proof of (2.17) in Lemma 2.14.

2nd proof of Theorem 2.20. Let $x_0 \in [0,1]$ be arbitrary but fixed. We shall show that

$$|u'(x_0)| \le C\left[1+\varepsilon^{-1}e^{-\alpha(1-x_0)/\varepsilon}\right].$$

If $x_0 \ge 1-\varepsilon$, then the result is immediate from Lemma 2.19, so we can assume that $0 \le x_0 \le 1-\varepsilon$. For $x \in [x_0,1]$, set $\psi(x) = u(x) - u(x_0)$,

$$C_5 = \frac{C_0 + C_1\|b\|_\infty}{\underline{a}}, \qquad C_6 = \frac{C_0}{\alpha(\underline{a}-\alpha)} + \frac{2C_1}{1-e^{-\alpha}},$$

and

$$\phi(x) = C_5(x-x_0) + C_6\left[e^{-\alpha(1-x)/\varepsilon} - e^{-\alpha(1-x_0)/\varepsilon}\right],$$

where C_0 and C_1 are defined in (2.19) and Lemma 2.17. We claim that ϕ is a barrier function for ψ on the interval $[x_0,1]$.

Now $|\psi(x_0)| = 0 = \phi(x_0)$ and Lemma 2.17 implies that $|\psi(1)| = |u(1)-u(x_0)| \le 2C_1 \le \phi(1)$ owing to the definition of C_6 and $1-x_0 \ge \varepsilon$. Furthermore, for $x \in (x_0,1)$ one has

$$\begin{aligned} |L\psi(x)| &= |L[u(x)-u(x_0)]| = |f(x,\varepsilon) - b(x)u(x_0)| \\ &\le C_0\left(1+\varepsilon^{-1}e^{-\alpha(1-x)/\varepsilon}\right) + C_1\|b\|_\infty \end{aligned} \tag{2.23}$$

by (2.19) and Lemma 2.17, while a short calculation shows that

$$\begin{aligned} L\phi(x) &= C_6\varepsilon^{-1}e^{-\alpha(1-x)/\varepsilon}\alpha\left[a(x)-\alpha\right] + C_5a(x) + b(x)\phi(x) \\ &\ge C_6\varepsilon^{-1}e^{-\alpha(1-x)/\varepsilon}\alpha\left[\underline{a}-\alpha\right] + C_5\underline{a}\,. \end{aligned}$$

Comparing this with (2.23), it is clear that the definitions of C_5 and C_6 imply that $L\phi(x) \ge |L\psi(x)|$. Thus ϕ is a barrier function for ψ on the interval $[x_0,1]$ and Corollary 1.12 yields $\phi(x) \ge |\psi(x)|$ on $[x_0,1]$.

Hence

$$\begin{aligned} |u'(x_0)| &= \left|\lim_{x\to x_0^+}\frac{\psi(x)}{x-x_0}\right| \\ &\le \lim_{x\to x_0^+}\left|\frac{\phi(x)}{x-x_0}\right| = |\phi'(x_0)| = C_5 + C_6\alpha\varepsilon^{-1}e^{-\alpha(1-x_0)/\varepsilon}, \end{aligned}$$

and we are done. (Note that one must have $\phi(x_0) = 0$ to derive the bound on $|u'(x_0)|$.) □

Remark 2.23. For the two-dimensional problem (2.22) it does not seem possible to generalize the above argument by finding a suitable barrier function that vanishes at the point (x_0, y_0) while satisfying all the inequalities required in the argument.

2.3.3. Using Green's function. Green's functions for (2.14) were discussed in section 2.2.

Andreev [**And02**] works with the special case $g_0 = g_1 = 0$. He derives various weighted estimates of Green's function $G(x, \xi)$ associated with (2.14) by considering it as a perturbation of Green's function for the case where $b \equiv 0$. (If $b \equiv 0$, then Green's function can be written down explicitly as a generalisation of (2.12).) He is thereby able to prove the inequalities

$$|u'(x)| \le C\left[1+\varepsilon^{-1}e^{-\alpha(1-x)/\varepsilon}\right]\|f\|_\infty \quad \forall x\in[0,1], \tag{2.24a}$$

$$\max_{0\le x\le 1}\left|(|u(x)|+\varepsilon|u'(x)|)e^{\alpha(1-x)/\varepsilon}\right| \le C\varepsilon \max_{0\le x\le 1}\left|f(x,\varepsilon)e^{\alpha(1-x)/\varepsilon}\right|. \tag{2.24b}$$

Since (2.19) gives only $\|f\|_\infty = \mathcal{O}\left(\varepsilon^{-1}\right)$, inequality (2.24a) does not provide an immediate proof of Theorem 2.20.

3rd proof of Theorem 2.20. By a change of dependent variable, we can assume that $g_0 = g_1 = 0$ without disturbing any of our hypotheses (the value of C_0 in (2.19) will then change, but we ignore this detail here). First decompose f into two components of distinct types: From (2.19) one sees that $|f(x)| \le 2C_0$ for $0 \le x \le 1-(\varepsilon/\alpha)|\ln\varepsilon|$. Choose $f_0 \in C[0,1]$ to agree with f on the interval $[0, 1-(\varepsilon/\alpha)|\ln\varepsilon|]$ and to satisfy $\|f_0\|_\infty \le 2C_0$. Set $f_1 = f - f_0$. Then $f_1 \equiv 0$ on $[0, 1-(\varepsilon/\alpha)|\ln\varepsilon|]$, while for $x \ge 1-(\varepsilon/\alpha)|\ln\varepsilon|$ one has

$$|f_1(x)| \le |f(x)| + |f_0(x)| \le C_0\left(3+\varepsilon^{-1}e^{-\alpha(1-x)/\varepsilon}\right) \le 4C_0\varepsilon^{-1}e^{-\alpha(1-x)/\varepsilon}.$$

For $i = 0, 1$, define $v_i \in C^2[0,1]$ to be the solution of $Lv_i = f_i$ on $(0,1)$ with $v_i(0) = v_i(1) = 0$. Applying (2.24a) to v_0 yields

$$|v_0'(x)| \le C\left[1+\varepsilon^{-1}e^{-\alpha(1-x)/\varepsilon}\right],$$

while applying (2.24b) to v_1 yields a similar result. But $u = v_0 + v_1$, so the proof is complete. □

Remark 2.24. As the Green's function for (2.22) is more complicated and less well-behaved than the Green's function for (2.14), it is uncertain whether an argument like this could work in the two-dimensional case.

2.3.4. Applying L to $u'(x)$ directly. The idea of this subsection is the most obvious one of all: one uses the barrier function technique of Corollary 1.12 to bound $u'(x)$ for $x \in [0,1]$. This technique has been used by many authors. To push through the argument, one needs the following extension of Corollary 1.12 to more general operators.

Lemma 2.25 (Barrier function without requiring $b \ge 0$). *Define the operator* $M : C^2(0,1) \to C(0,1)$ *by*

$$Mv(x) := -\varepsilon v''(x) + a(x)v'(x) + \tilde{b}(x)v(x) \quad \forall x \in (0,1),$$

where $\tilde{b} \in C[0,1]$ *satisfies* $\underline{a}^2 + 4\varepsilon\tilde{b}(x) \geq 0$ *for all* x *(here* $\tilde{b} < 0$ *is permitted). Let* $v, w \in C^2(0,1) \cap C[0,1]$ *satisfy* $Mv(x) \geq |Mw(x)|$ *on* $(0,1)$ *and* $v(x) \geq |w(x)|$ *for* $x = 0, 1$. *Then* $v \geq |w|$ *on* $[0,1]$.

Proof. Set $w(x) = e^{\sigma x}\tilde{w}(x)$ for $x \in [0,1]$, where σ is independent of x and will be specified in a moment. Then a calculation gives

$$\begin{aligned} Mw(x) &= e^{\sigma x}\left\{-\varepsilon\tilde{w}''(x) + [a(x) - 2\varepsilon\sigma]\,\tilde{w}'(x) + \left[\tilde{b}(x) + a(x)\sigma - \varepsilon\sigma^2\right]\tilde{w}(x)\right\} \\ &= e^{\sigma x}\tilde{M}\tilde{w}(x), \end{aligned}$$

say. Similarly, setting $v(x) = e^{\sigma x}\tilde{v}(x)$, one gets $Mv(x) = e^{\sigma x}\tilde{M}\tilde{v}(x)$, so we now have $\tilde{M}\tilde{v}(x) \geq |\tilde{M}\tilde{w}(x)|$ on $(0,1)$. Moreover, $\tilde{v}(x) \geq |\tilde{w}(x)|$ for $x = 0, 1$. Set $\underline{\tilde{b}} = \min_{0\leq x\leq 1}\tilde{b}(x)$. Choose $\sigma = \left[\underline{a} + \sqrt{\underline{a}^2 + 4\varepsilon\underline{\tilde{b}}}\right]/(2\varepsilon)$. Then $0 < \sigma$ and $-\varepsilon\sigma^2 + \underline{a}\sigma + \underline{\tilde{b}} = 0$. Thus $\tilde{b}(x) + a(x)\sigma - \varepsilon\sigma^2 \geq 0$; consequently, $\tilde{M}$ satisfies the comparison principle of Corollary 1.12. Hence $\tilde{v}(x) \geq |\tilde{w}(x)|$ on $[0,1]$, which gives $v(x) \geq |w(x)|$ on $[0,1]$, as desired. □

Exercise 2.26. Show clearly the connection between this lemma and Remark 2.12.

Note that the hypothesis $\underline{a}^2 + 4\varepsilon\tilde{b} \geq 0$ of Lemma 2.25 will be satisfied automatically for all sufficiently small ε, irrespective of the sign of $\tilde{b}$. Variants of this lemma have been used by several authors; the earliest example seems to be [**Lor81**].

Assume that $a, b \in C^1[0,1]$.

4th proof of Theorem 2.20. From Lemmas 2.17 and 2.19 one has $|u'(0)| \leq C_2$ and $|u'(1)| \leq C_4\varepsilon^{-1}$. Now

$$\begin{aligned} L(u') &= -\varepsilon u''' + au'' + bu' \\ &= (-\varepsilon u'' + au' + bu)' - a'u' - b'u \qquad (2.25) \\ &= f_x - a'u' - b'u. \end{aligned}$$

Define the operator $\hat{L} : C^2[0,1] \to C(0,1)$ by $\hat{L}v = Lv + a'v$. Then $\hat{L}(u') = f_x - b'u$. Hence

$$|\hat{L}(u'(x))| \leq C_7\left(1 + \varepsilon^{-2}e^{-\alpha(1-x)/\varepsilon}\right) \quad \text{for all } x \in (0,1),$$

where C_7 is some constant. We shall apply the comparison principle of Lemma 2.25 to $\hat{L}$ and the function u', by constructing a suitable barrier function. For any constant k one has $\hat{L}(e^{kx}) = e^{kx}(-\varepsilon k^2 + ak + b + a')$; choosing $k = 2(\|b\|_\infty + \|a'\|_\infty)/\alpha$ and taking ε sufficiently small yields

$\hat{L}(e^{kx}) \geq C_8 := \|b\|_\infty + \|a'\|_\infty$ for all $x \in [0,1]$. One also has

$$\hat{L}\big(e^{-\alpha(1-x)/\varepsilon}\big) = \left\{\frac{\alpha[a(x)-\alpha]}{\varepsilon} + b(x) + a'(x)\right\} e^{-\alpha(1-x)/\varepsilon}$$
$$\geq \left\{\frac{\alpha\,[\underline{a}-\alpha]}{\varepsilon} - \|b\|_\infty - \|a'\|_\infty\right\} e^{-\alpha(1-x)/\varepsilon}.$$

Thus for ε sufficiently small one obtains $\hat{L}\big(e^{-\alpha(1-x)/\varepsilon}\big) \geq C_9\varepsilon^{-1}e^{-\alpha(1-x)/\varepsilon}$ for some positive constant C_9 and all x. These inequalities together yield

$$\hat{L}\big(C_7(e^{kx}/C_8 + \varepsilon^{-1}e^{-\alpha(1-x)/\varepsilon}/C_9)\big) \geq |\hat{L}u'(x)| \quad \text{for all } x \in (0,1).$$

After modifying the constants in the barrier function to handle the boundary data for u', Lemma 2.25 then gives

$$|u'(x)| \leq \left(C_2 + \frac{C_7}{C_8}\right)e^{kx} + \left(C_2 + \frac{C_7}{C_9}\right)\varepsilon^{-1}e^{-\alpha(1-x)/\varepsilon} \quad \text{for all } x \in (0,1). \qquad \square$$

2.3.5. Bounds on higher-order derivatives of u. The next theorem gives bounds on all derivatives of u. These bounds are sharp in their powers of ε and the exponential decay away from $x = 1$; recall Example 1.1.

Theorem 2.27. *Assume that the functions a and b are smooth, and assume that f satisfies* (2.19). *Then there exists a constant $C = C(\alpha, q)$ such that for all $x \in [0,1]$ the solution of* (2.14) *satisfies*

$$|u^{(i)}(x)| \leq C\left(1 + \varepsilon^{-i}e^{-\alpha(1-x)/\varepsilon}\right) \quad \text{for } i = 0,1,2,\dots,q. \tag{2.26}$$

Proof. The case $i = 0$ is proved in Lemma 2.17 and this is used in section 2.3.1 to prove the case $i = 1$. One can modify these arguments to prove Theorem 2.27 by induction on i. See Exercise 2.28 or [**KT78**][1] for more details. $\square$

Theorem 2.27 implies that $|u^{(k)}(1)| = \mathcal{O}\left(\varepsilon^{-k}\right)$ for $k = 1,2,\dots$.

Exercise 2.28. We say that a function $g(x,\varepsilon)$ is of class j if for some constant C and all $x \in [0,1]$ one has

$$\left|\frac{\partial^i g(x,\varepsilon)}{\partial x^i}\right| \leq C\left(1 + \varepsilon^{-(i+1)}e^{-\alpha(1-x)/\varepsilon}\right)$$

for $i = 0,1,2,\dots,j$. Consider the two-point boundary value problem

$$Ly(x) = g(x,\varepsilon) \text{ on } (0,1), \quad y(0) = y_0,\ y(1) = y_1,$$

[1] *Historical Note.* The first proof of Theorem 2.27 was given in a famous paper by Kellogg and Tsan [**KT78**]. Subsequently, Bruce Kellogg wrote many papers on convection-diffusion problems, but this highly cited paper was the only mathematical paper that Alice Tsan ever published!

where g is of class j and y_0, y_1 are constants. Use induction on j to prove that for some constant C (independent of ε) one has

$$|y^{(i)}(x)| \le C\left[1 + \varepsilon^{-i}e^{-\alpha(1-x)/\varepsilon}\right] \qquad \text{for } x \in [0,1] \text{ and } i = 0, 1, \dots, j+1.$$

Hint. For the inductive step, suppose that the result is true for $j = k$. Differentiate $k+1$ times the equation $Ly = g$, and set $z = y^{(k+1)}$. Then $-\varepsilon z'' + az' = r$, where r depends on y, a, b, g and their derivatives of order at most $k+1$. By the inductive hypothesis,

$$|r(x)| \le C\left[1 + \varepsilon^{-(k+2)}\exp(-\alpha(1-x)/\varepsilon)\right].$$

Now imitate the proof of Lemma 2.19 to bound $y^{(k+2)}(1)$, then mimic the analysis of section 2.3.1 to complete the inductive step.

Remark 2.29. In Exercise 2.28, note that near $x = 1$ each derivative $y^{(i)}$ is better behaved than $g^{(i)}$ by an extra factor ε. This *smoothing property* of the differential operator L will be exploited later to prove Theorem 3.11.

Remark 2.30. As well as the pointwise bounds of Theorem 2.27, estimates of the solution u of (2.14) can be derived in other norms; see [**Lin10**, Theorem 3.25] and [**RST08**, Theorem I.1.7].

Exercise 2.31. When discussing finite element methods, we shall use the standard Sobolev spaces H^k with their associated norms $\|\cdot\|_k$; in particular $\|\cdot\|_0 = \|\cdot\|_{L_2}$. Use Theorem 2.27 to show that

$$\|u\|_k \le \begin{cases} C & \text{if } k = 0, \\ C\varepsilon^{-k+\frac{1}{2}} & \text{if } k = 1, 2, \dots, q. \end{cases}$$

A more direct derivation of Sobolev-norm a priori bounds on solutions of convection-diffusion problems will be given in Lemma 4.14 and section 6.1.

Exercise 2.32. If $f \equiv 0$ and $g_0 = 0$ in (2.14), prove that the solution u is a pure boundary layer, viz., $|u'(x)| \le C\varepsilon^{-1}e^{-\alpha(1-x)/\varepsilon}$ for $0 \le x \le 1$. *Hint.* Bound $u(x)$ using the barrier function $|g_1|e^{-\underline{a}(1-x)/\varepsilon}$, then invoke this bound in section 2.3.1.

Remark 2.33. If in (2.14) we replace the Dirichlet boundary condition $u(1) = g_1$ at the layer by a *Neumann boundary condition* $u'(1) = k$ (for some constant k), then it turns out that (2.26) can be replaced by

$$|u^{(i)}(x)| \le C\left(1 + \varepsilon^{1-i}e^{-\alpha(1-x)/\varepsilon}\right) \qquad \text{for } i = 0, 1, 2, \dots, q. \tag{2.27}$$

That is, the first-order derivative of u is bounded at $x = 1$ as $\varepsilon \to 0$ (this is obvious a priori from the Neumann boundary condition), while the higher-order derivatives of u at $x = 1$ still blow up as $\varepsilon \to 0$, but not as badly as

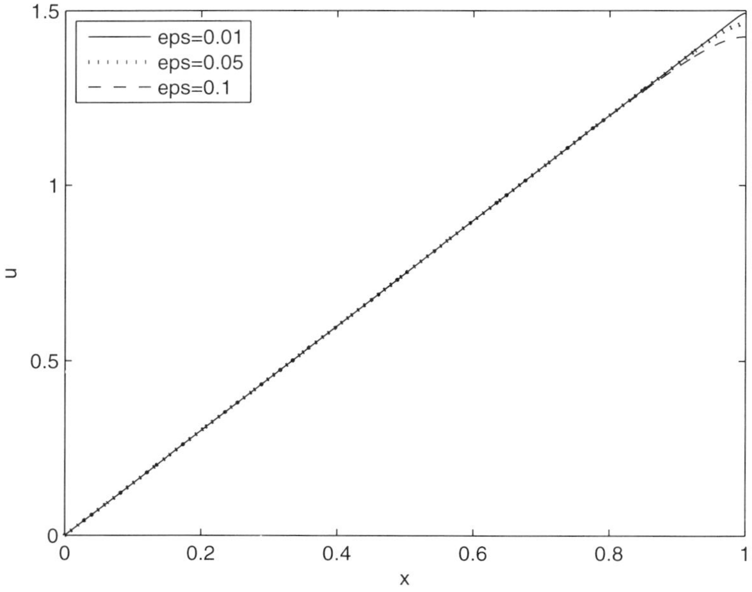

Figure 2.3. Graph of solution to Neumann problem (2.28) with $\varepsilon = 0.1, 0.05, 0.01$

for a Dirichlet boundary condition. Figure 2.3 displays the solution for the example

$$-\varepsilon u''(x) + 2u'(x) = 3 \quad \text{for } 0 < x < 1, \tag{2.28a}$$

$$u(0) = 0, \; u'(1) = 0, \tag{2.28b}$$

where ε takes the values 0.01, 0.05, and 0.1. Here Example 1.1 has been modified by replacing the Dirichlet condition $u(1) = 0$ by the Neumann condition $u'(1) = 0$.

Figure 2.4 is a zoom of Figure 2.3 near $x = 1$. These two figures display no obvious layer in $u(x)$ at $x = 1$, but the function is nevertheless not entirely well-behaved.

Exercise 2.34. Compute the exact solution of the boundary value problem (2.28). Hence show that the first-order derivative of u is bounded at $x = 1$ as $\varepsilon \to 0$, but the higher-order derivatives of u at $x = 1$ will blow up as $\varepsilon \to 0$. Verify that this function does not satisfy the $L^\infty[0,1]$ definition of "singularly perturbed" that is stated in (2.2), despite the bad behaviour of its higher-order derivatives.

There are alternative definitions of "singularly perturbed" that will be satisfied by this function, e.g., replace u in (2.2) by u'; this is equivalent to replacing $L^\infty[0,1]$ by the stronger Sobolev norm $W^{1,\infty}[0,1]$.

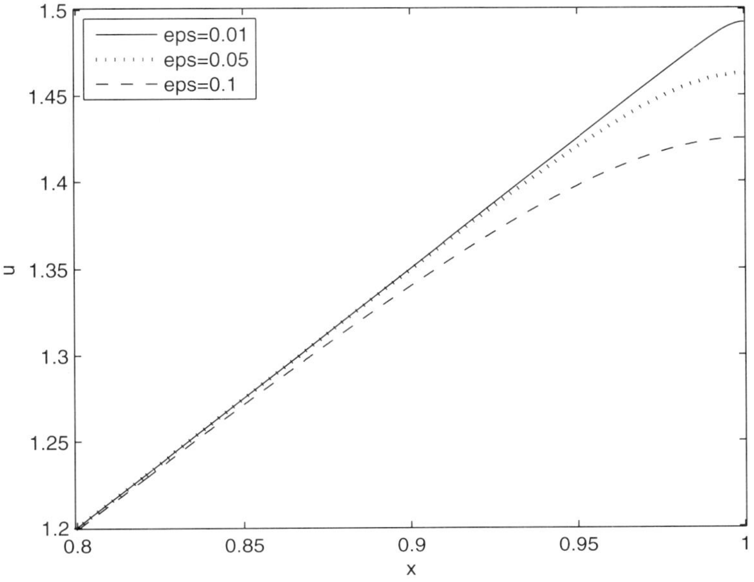

Figure 2.4. Zoom of solution to Neumann problem (2.28) near $x = 1$ with $\varepsilon = 0.1, 0.05, 0.01$

Exercise 2.35. Prove the bound (2.27) of Remark 2.33. *Hint.* Use Exercise 1.11 to bound $\|u\|_\infty$ and $|u'(0)|$. Find a boundary value problem satisfied by $u'(x)$, then apply Theorem 2.27 to this problem to bound $|u^{(i)}(x)|$ for $i > 0$. For convenience assume in the analysis that $b + a' > 0$, though this assumption could be avoided by a change of variable, as in Remark 2.12.

Exercise 2.36. Let u be the solution of the convection-diffusion problem (2.14). Suppose that the solution u_0 of the reduced problem (2.18) happens to satisfy the boundary condition $u_0(1) = g_1$. Prove that

$$|u^{(i)}(1)| \le C\varepsilon^{1-i} \quad \text{for } i = 1, 2, \ldots, q-1,$$

for all $x \in [0, 1]$ and some constant C; thus the solution of this problem is better behaved than typical solutions of (2.14). *Hint.* What boundary value problem is satisfied by the function $u - u_0$?

One might ask, when u is the solution of the convection-diffusion problem (2.14), can't we obtain bounds on derivatives of u simply by differentiating uniform asymptotic expansions such as (2.8)? This is tempting, but we have developed no theory that controls the difference between a derivative of u and the same derivative of its asymptotic expansion. In general the differentiation of asymptotic expansions of functions is not rigorously justified, but for solutions of elliptic differential equations, a theory can be established.

This approach is outlined in Theorem 2.44 below, and it leads not only to bounds on the derivatives of u but also to a convenient decomposition of u.

Remark 2.37. Singularly perturbed linear *reaction-diffusion* problems, as in Exercise 2.8, have the form

$$-\varepsilon u''(x) + b(x)u(x) = r(x) \text{ on } (0,1), \quad u(0) = \gamma_0, \ u(1) = \gamma_1, \tag{2.29}$$

where γ_0 and γ_1 are given constants, $b \geq \beta^2$ for some positive constant β, and $b, r \in C^q[0,1]$ for some $q \geq 1$. In reaction-diffusion problems there is no convection term—a significant change from our previous theory.

For the boundary value problem (2.29), the reduced solution u_0 is obtained—as before—by setting $\varepsilon = 0$; this yields $bu_0 = r$, i.e., $u_0(x) = r(x)/b(x)$. Note that no boundary condition is needed for this reduced problem! Away from $x = 0$ and $x = 1$, the reduced solution is an accurate approximation of the true solution u of (2.29), but u typically has boundary layers at $x = 0$ and $x = 1$ because at these points the reduced solution fails usually to satisfy the prescribed boundary condition.

The derivatives of u satisfy

$$|u^{(i)}(x)| \leq C\left(1 + \varepsilon^{-i/2}e^{-\beta x/\sqrt{\varepsilon}} + \varepsilon^{-i/2}e^{-\beta(1-x)/\sqrt{\varepsilon}}\right) \tag{2.30}$$

for $i = 0, 1, \ldots, q$ and $0 \leq x \leq 1$. Note that the scaling of the layers is $1/\sqrt{\varepsilon}$, unlike the $1/\varepsilon$ scaling of convection-diffusion problems.

Exercise 2.38. Let u be the solution of the reaction-diffusion problem (2.29). Prove that there exists a constant C such that $\|u'\|_\infty \leq C\varepsilon^{-1/2}$ by modifying the proof of Lemma 2.19.

Exercise 2.39. State and prove an analogue of Exercise 2.16 for the reaction-diffusion problem (2.29).

Exercise 2.40. Let u be the solution of the reaction-diffusion problem (2.29). Suppose that $\gamma_1 b(1) = r(1)$, i.e., the boundary condition at $x = 1$ happens to satisfy the reduced problem. Prove that $|u'(1)| \leq C$ for some constant C. *Hint.* Use a barrier function, much like the proof of Lemma 2.14.

The best source of information about singularly perturbed reaction-diffusion problems is Linß's book [**Lin10**]; see also [**RST08**].

Example 2.41. Consider the reaction-diffusion problem

$$-\varepsilon u''(x) + 2u(x) = 3 \quad \text{for } 0 < x < 1, \tag{2.31a}$$

$$u(0) = u(1) = 0. \tag{2.31b}$$

Here we have modified Example 1.1 by changing $u'(x)$ to $u(x)$. One can easily compute $u(x)$ explicitly (do this as an exercise to see how the bounds in (2.30) arise), and its graph is drawn in Figure 2.5. Note that $u(x)$ has

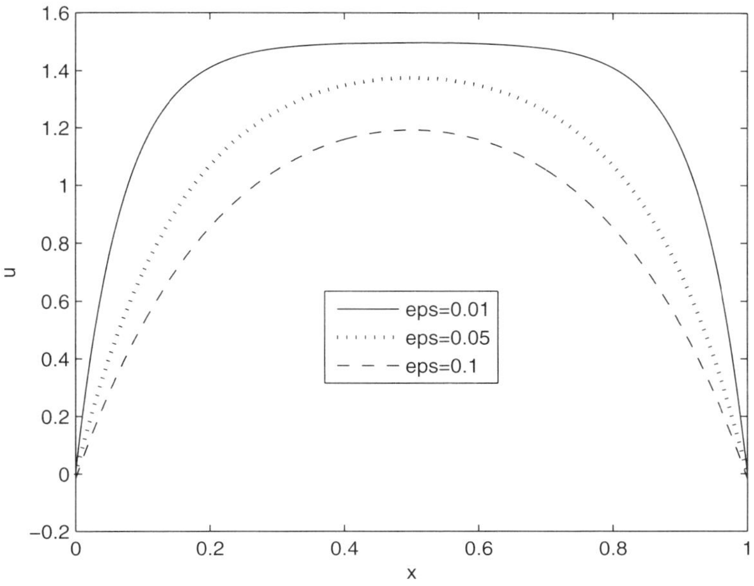

Figure 2.5. Solution to reaction-diffusion problem (2.31) with $\varepsilon = 0.1, 0.05, 0.01$

boundary layers at both $x = 0$ and $x = 1$ and these layers are less sharp than in the convection-diffusion example of Figure 1.1 because u' is $\mathcal{O}\left(1/\sqrt{\varepsilon}\right)$ in reaction-diffusion boundary layers but u' is $\mathcal{O}\left(1/\varepsilon\right)$ in convection-diffusion boundary layers.

Remark 2.42. If $a(x)$ changes sign inside the domain (the point where this happens is called a *turning point*), then the solution $u(x)$ may have interior layers and/or boundary layers. The simplest case is when $a(x) = (x - x_0)\hat{a}(x)$ for some $x_0 \in (0, 1)$ and $\hat{a} > 0$ on $[0, 1]$, with $b > 0$ on $[0, 1]$; then $a > 0$ on $(x_0, 1]$ and our usual exponential boundary layer appears at $x = 1$, while $a < 0$ on $[0, x_0)$ causes another exponential boundary layer at $x = 0$ (recall Exercise 2.10). There are no other layers.

If $a(x) = -(x - x_0)\hat{a}(x)$ with $\hat{a} > 0$ on $[0, 1]$, the solution is entirely different: it has an interior layer at $x = x_0$ and no boundary layers. For example, the solution of

$$-\varepsilon u''(x) - 2xu'(x) = 2\exp(-x^2/\varepsilon) \quad \text{on } (-1, 1), \tag{2.32a}$$

$$u(-1) = u(1) = 0 \tag{2.32b}$$

is $u(x) = \exp(-x^2/\varepsilon) - \exp(-1/\varepsilon)$; its graph is drawn in Figure 2.6.

Interior layers caused by turning points have a more complicated structure than exponential boundary layers, and we do not discuss them further;

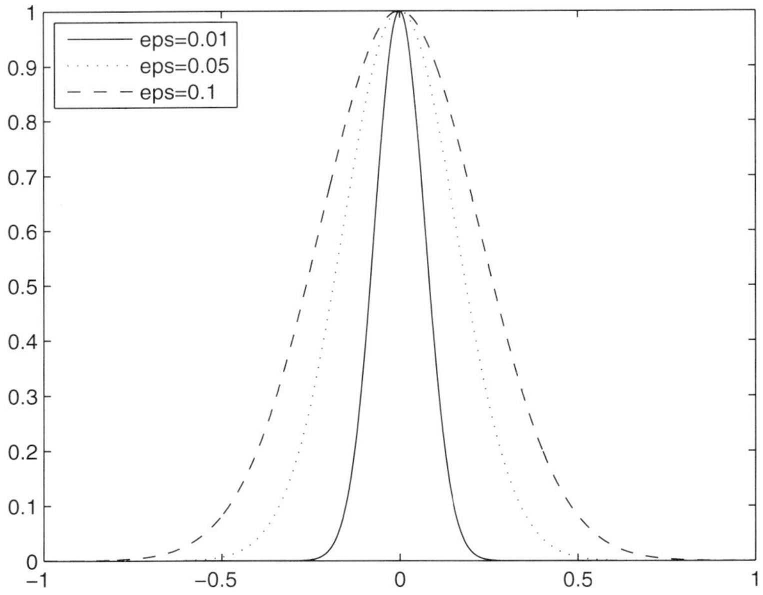

Figure 2.6. Interior layer: solution of (2.32) with $\varepsilon = 0.1, 0.05, 0.01$

see [**Lin10**, Section 3.5] and [**RST08**, Section I.1.2], and a survey of the literature on turning point problems is given in [**SRP13**].

Exercise 2.43. An interior layer can also arise if f is discontinuous. These layers are simpler than the turning point interior layers of Remark 2.42. Compute the exact solution of the problem

$$-\varepsilon u'' + 2u' = f \text{ on } (0,2), \quad u(0) = u(2) = 0,$$

where $f = 1$ on $(0,1)$ and $f = 3$ on $(1,2)$, and hence write down bounds on the derivatives of u.

2.4. Decompositions of the solution

Theorems 2.44 and 2.48 will show that the solution u of the convection-diffusion problem (2.14) can be written as the sum of a well-behaved term and a layer term. Such decompositions of u aid our insight when constructing accurate numerical methods and are often needed in the rigorous analysis of such methods.

Theorem 2.44 (Standard decomposition of u)**.** *Let q be a positive integer. Let u be the solution of (2.14). Assume that the functions a, b, and f are sufficiently smooth. Then there is a splitting $u = S + E$ such that*

$$\|S^{(j)}\|_\infty \le C \quad \text{and} \quad |E^{(j)}(x)| \le C\varepsilon^{-j}e^{-\alpha(1-x)/\varepsilon}$$

for $0 \le j \le q$ and $0 \le x \le 1$, where the constant $C = C(q)$.

Proof. Recall the standard asymptotic expansion of $u(x)$ given in (2.8), and for convenience write $R(x)$ for the remainder $R(x, \varepsilon, k)$. Observe that we have a bound only on $\|R\|_\infty$; no information is available on the derivatives of $R(x)$. As the u_n and v_n are computed explicitly and $Lu = f$, one can determine $LR(x)$ from (2.8). Now the deep a priori estimates of Schauder for elliptic differential equations [**LU68**, p. 110] will yield the bound $\|R^{(j)}\|_\infty \le C\varepsilon^{-j}$ for $0 \le j \le q$ (one first needs to stretch the variable x via the transformation $x \to t := x/\varepsilon$ to obtain a standard elliptic operator to which [**LU68**] can be applied).

Choosing $k = q-1$ in (2.8), set $S = \sum_{n=0}^{q-1} u_n(x)\varepsilon^n + \varepsilon^q R(x)$ and $E(x) = \sum_{n=0}^{q-1} v_n(x)\varepsilon^n$. The result now follows immediately from what is known about the terms in S and E. $\square$

In this theorem and other similar results, S is called the *smooth component* or *regular component* of u, and E is called the *layer component.*

Remark 2.45. In the literature dealing with singularly perturbed differential equations, "smooth" is generally used in this nonstandard way to mean that a function has certain low-order derivatives bounded *independently of the perturbation parameter.*

Theorem 2.27 is adequate when proving convergence of some numerical methods for (2.14), but for others one needs to invoke Theorem 2.44 in order to analyse separately the smooth and layer components of u. At first sight Theorem 2.44 seems the stronger of the two results, but this is not the case, as a result of Linß [**Lin01**] shows.[2]

Theorem 2.46. *Theorems* 2.27 *and* 2.44 *are equivalent.*

Proof. Clearly, Theorem 2.44 implies Theorem 2.27.

For the converse implication, assume that (2.26) holds true for some fixed positive integer q. Set $x^* = 1 - (q\varepsilon/\alpha)\ln(1/\varepsilon)$, and define $S(x) = u(x)$ for $0 \le x \le x^*$. Then (2.26) and the choice of x^* ensure that $|S^{(j)}(x)| \le C$ for $0 \le j \le q$ and $0 \le x \le x^*$. Consequently, one can (e.g., using a Taylor expansion of $S(x)$ about $x = x^*$) extend S to $[0,1]$ with $|S^{(j)}(x)| \le C$ for $0 \le j \le q$ and $0 \le x \le 1$.

Now set $E = u - S$. Then $E(x) \equiv 0$ for $0 \le x \le x^*$, and for $x^* < x \le 1$, we have

$$|E^{(q)}(x)| \le |u^{(q)}(x)| + |S^{(q)}(x)| \le C\left(1 + \varepsilon^{-q} e^{-\alpha(1-x)/\varepsilon}\right) \le C\varepsilon^{-q} e^{-\alpha(1-x)/\varepsilon}$$

[2] *Historical Note.* The author of the proof of Theorem 2.46, which appeared in [**Lin01**], is unknown! Torsten Linß submitted his paper [**Lin01**] to a journal, and received a referee report on the paper which stated and proved Theorem 2.46—but referee reports are always written anonymously so we do not know who this mysterious mathematician was.

from the definition of x^*. Using induction (with a decreasing index), integrate $E^{(k)}(x)$ for $k = q, q-1, \dots, 1$ to get

$$\begin{aligned} |E^{(k-1)}(x)| &= \left| \int_{x^*}^{x} E^{(k)}(s)\,\mathrm{d}s \right| \\ &\le C \int_{x^*}^{x} \varepsilon^{-k} e^{-\alpha(1-s)/\varepsilon}\,\mathrm{d}s \\ &\le C\varepsilon^{-(k-1)} e^{-\alpha(1-x)/\varepsilon} \quad \text{for } x^* < x \le 1. \end{aligned}$$ □

Exercise 2.47. Prove rigorously the claim in the above proof that if the function S is defined on $[0, x^*]$ with $|S^{(j)}(x)| \le C$ for $0 \le j \le q$ and $0 \le x \le x^* < 1$, then one can extend S to $[0, 1]$ with $|S^{(j)}(x)| \le C$ for $0 \le j \le q$ and $0 \le x \le 1$. (*Note.* The two constants "C" here can take different values.)

For the analysis of certain finite difference methods on Shishkin meshes (which we will meet in Section 3.4), one needs a decomposition of u with a further property that is originally due to Shishkin; see the references in the books [**FHM+00**] and [**MOS12**] by Shishkin et al. By slightly modifying the construction of the asymptotic expansion (2.8) (see [**DR97, MOS12**] and [**RST08**, page 23]), one can prove the following strengthening of Theorem 2.44:

Theorem 2.48 (Shishkin decomposition of u). *Let q be a positive integer. Let u be the solution of* (2.14). *Assume that the functions a, b, and f are sufficiently smooth. Then there is a splitting $u = S + E$ such that*

$$\|S^{(j)}\|_\infty \le C \text{ and } |E^{(j)}(x)| \le C\varepsilon^{-j} e^{-\alpha(1-x)/\varepsilon} \quad \text{for } 0 \le x \le 1 \tag{2.33}$$

for $0 \le j \le q$ and $0 \le x \le 1$, where the constant $C = C(q)$, and in addition

$$LS(x) = f(x) \text{ and } LE(x) = 0 \quad \text{for } 0 \le x \le 1.$$

Proof. In the standard asymptotic expansion for $Lu := \varepsilon u'' + au' + bu = f$, one has

$$u(x) = \sum_{n=0}^{q-1} \varepsilon^n u_n(x) + \sum_{n=0}^{q-1} \varepsilon^n v_n(\rho) + \varepsilon^q R(x, \varepsilon, q-1)$$

for each positive integer q, where $\rho = (1-x)/\varepsilon$. The terms u_n here are defined by

$$L_0 u_0 = f,\ u_0(0) = 0, \text{ and } L_0 u_n = -u''_{n-1},\ u_n(0) = 0 \text{ for } n = 1, 2, \dots, q-1,$$

where $L_0 : w \mapsto aw' + bw$ is the reduced operator obtained by setting $\varepsilon = 0$ in L. The $v_n(x)$ satisfy

$$\begin{aligned} &\tilde{L}v_0 = 0,\ v_0(0) = -u_0(1),\ v_0(\infty) = 0, \\ &\tilde{L}v_n = \tilde{L}^*(v_0, v_1, \dots, v_{n-1}),\ v_n(0) = -u_n(1), \\ &\qquad\qquad v_n(\infty) = 0 \text{ for } n = 0, 1, \dots, q-1, \end{aligned}$$

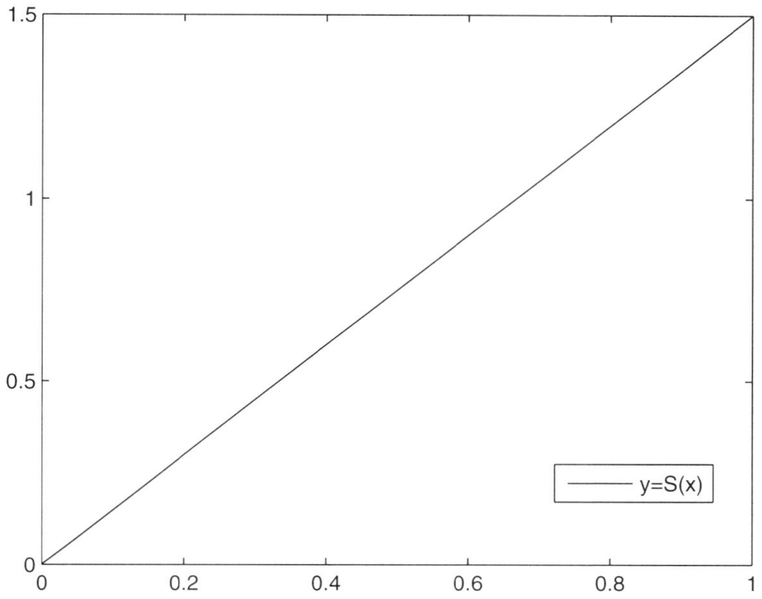

Figure 2.7. Graph of S for Example 1.1 when $\varepsilon = 0.01$

where $\tilde{L} = -\dfrac{d^2}{d\rho^2} + a(1)\dfrac{d}{d\rho}$ and $\tilde{L}^*(v_0, v_1, \ldots, v_{n-1}) = \displaystyle\sum_{i=1}^{n} \frac{a^{(i)}(1)}{i!}\frac{dv_{n-i}}{d\rho}$.

The Shishkin decomposition is

$$u(x) = \underbrace{\sum_{n=0}^{q-1} \varepsilon^n u_n(x) + \varepsilon^q u_q^*}_{=:S} + \underbrace{\sum_{n=0}^{q-1} \varepsilon^n v_n(\rho) + \varepsilon^q v_q^*}_{=:E},$$

where

$$Lu_q^* = u_{q-2}'', \ u_q(0) = u_q(1) = 0$$

(note that L, not L_0, is used here) and

$$\tilde{L}v_q^* = -\varepsilon^{-q}\tilde{L}\left(\sum_{n=0}^{q-1} v_n(\rho)\right), \ v_q^*(0) = 0, \ v_q^*(1/\varepsilon) = -\sum_{n=0}^{q-1} v_n(1/\varepsilon).$$

One can use Theorem 2.27 to verify that the conclusions of Theorem 2.48 apply to S. To bound E and its derivatives, use Exercise 2.32. □

Graphs of S and E for Example 1.1 are displayed in Figures 2.7 and 2.8.

Remark 2.49. The history of solution decompositions such as those described in Theorems 2.44 and 2.48 can be traced back to earlier work of Bakhvalov and Volkov; see [**KO10**].

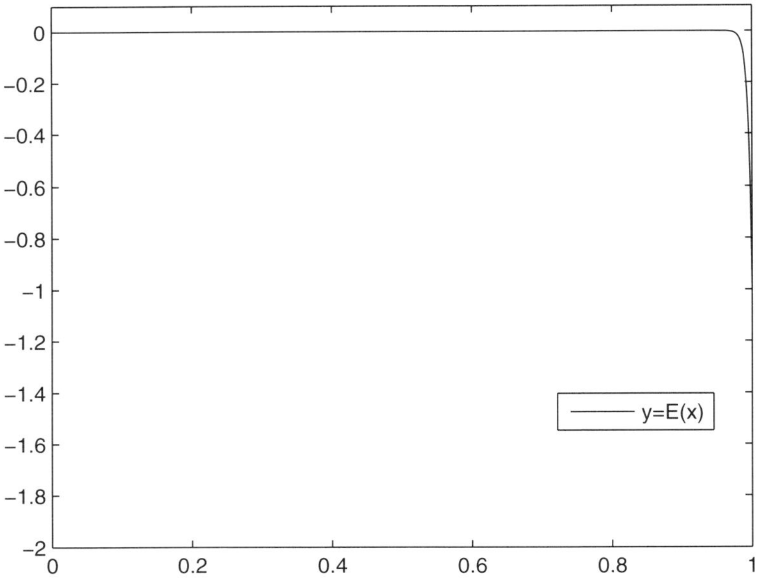

Figure 2.8. Graphs of E for Example 1.1 when $\varepsilon = 0.01$

Remark 2.50. Reaction-diffusion problems of the form $Lu = -\varepsilon u'' + bu = r$ on $(0,1)$, with $u(0)$ and $u(1)$ given and $b \geq \beta^2 > 0$ on $[0,1]$, were discussed in Remark 2.37. The solutions u of such problems have layers at both $x = 0$ and $x = 1$, and the decomposition that is the analogue of Theorem 2.48 is derived in [**MOS12**, Chapter 6]: $u = S + E_0 + E_1$, where

$$\|S^{(j)}\|_\infty \leq C,\ |E_0^{(j)}(x)| \leq C\varepsilon^{-j/2}e^{-\beta x/\sqrt{\varepsilon}}, |E_1^{(j)}(x)| \leq C\varepsilon^{-j/2}e^{-\beta(1-x)/\sqrt{\varepsilon}}$$

for $0 \leq j \leq q$ and $0 \leq x \leq 1$, with

$$LS(x) = r(x),\ LE_0(x) = LE_1(x) = 0 \quad \text{for } 0 \leq x \leq 1.$$

Here q is a positive integer, $C = C(q)$, and we assume that the functions b and r are sufficiently smooth.

Chapter 3

Finite Difference Methods in One Dimension

Consider the convection-diffusion problem

(3.1a) $Lu(x) := -\varepsilon u''(x) + a(x)u'(x) + b(x)u(x) = f(x) \quad \text{for } 0 < x < 1,$

(3.1b) $u(0) = u(1) = 0,$

where $0 < \varepsilon \leq 1$, $a(x) \geq \underline{a} > \alpha > 0$, and $b(x) \geq 0$ on [0,1]. Assume that a, b, and f lie in $C^1[0,1]$. As we pointed out in section 2.1, general Dirichlet boundary conditions $u(0) = g_0, u(1) = g_1$ are easily reduced to (3.1b) by a simple change of variable.

Let N be a positive integer. We partition [0,1] by the equidistant mesh $x_i = ih$ for $i = 0, \ldots, N$, where $h := 1/N$. On this mesh we examine how to compute an approximation $\vec{u}^N := (u_0^N \; u_1^N \cdots u_N^N)^T$ of $(u_0 \; u_1 \cdots u_N)^T$, where T denotes transpose; here and subsequently we write u_i for $u(x_i)$, a_i for $a(x_i)$, etc.

Standard discretizations of differential equations use a *central difference approximation* of the convective term. That is, one approximates $u'(x_i)$ by $(u_{i+1}^N - u_{i-1}^N)/(2h)$, which is formally an $\mathcal{O}\left(h^2\right)$ approximation. Using this discretization and the standard approximation $(u_{i-1}^N - 2u_i^N + u_{i+1}^N)/h^2$ of $u''(x_i)$ produces a difference scheme $B\vec{u}^N = \vec{f}^N$ whose matrix B is tridiagonal with ith row

(3.2) $$\left(0 \cdots 0 \quad -\frac{\varepsilon}{h^2} - \frac{a_i}{2h} \quad \frac{2\varepsilon}{h^2} + b_i \quad -\frac{\varepsilon}{h^2} + \frac{a_i}{2h} \quad 0 \cdots 0\right)$$

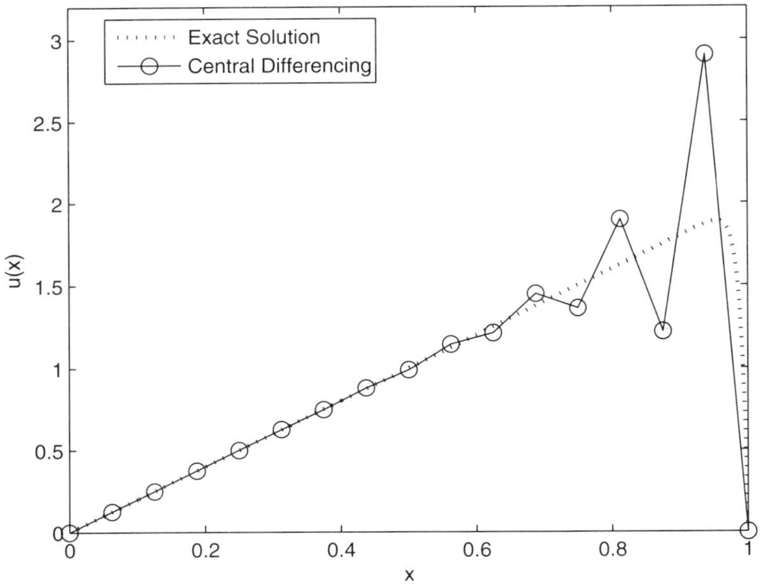

Figure 3.1. Example 1.1 with $\varepsilon = 0.01$; solution computed by central differencing with $N = 16$

for $i = 1, \dots, N-1$. The 0th and Nth rows of B, which impose the boundary conditions (3.1b), are $(1\ 0 \cdots 0)$ and $(0 \cdots 0\ 1)$. The right-hand side of the scheme is $\vec{f}^N := (0\ f_1\ f_2 \cdots f_{N-1}\ 0)^T$.

In the particular case of Example 1.1, where $a(x) \equiv 2, b(x) \equiv 0$ and $f(x) \equiv 3$, the solution of this difference scheme is

$$u_i^N = \frac{3x_i}{2} - \frac{3(r^{N-i} - r^N)}{2(1 - r^N)} \qquad \text{with } r := \frac{2\varepsilon - h}{2\varepsilon + h}$$

(we leave this as an exercise). In practice one usually has $N \ll 1/\varepsilon$, so $\varepsilon \ll h$ and $r \approx -1$. Consequently, the computed solution will oscillate as i varies, quite unlike the true solution (1.2); see Figure 3.1.

Remark 3.1. To see that these oscillations are also present in the general case (3.1), consider (3.2) with $i = N - 1$. Taking $\varepsilon \ll h^2$, this equation is essentially

$$f_{N-1} = \frac{a_{N-1}(u_N^N - u_{N-2}^N)}{2h} + b_{N-1}u_{N-1}^N = -\frac{a_{N-1}u_{N-2}^N}{2h} + b_{N-1}u_{N-1}^N,$$

upon applying the boundary condition. Hence, $u_{N-2}^N = \mathcal{O}(h)$; but because of the boundary layer in $u(x)$ at $x = 1$, we expect that u_{N-2} is not close to zero. Thus u_{N-2}^N is far from the true value u_{N-2}, and this is due to oscillations in the computed solution.

Exercise 3.2. Suppose that central differencing on an equidistant mesh of width $h = 1/N$ is used to solve the problem

$$-\varepsilon u'' + au' + bu = 0 \text{ on } (0,1), \quad u(0) = 0,\ u(1) = 1,$$

where $a > 0$ and $b \geq 0$. Assume that $\min_x a(x) > 2\varepsilon/h$. By considering the signs of the coefficients in the difference scheme, show that the computed solution $\{u_i^N\}_{i=0}^N$ has $u_i^N u_{i+1}^N < 0$ for $i = 1, 2, \ldots, N-1$. This conclusion says that the computed solution oscillates around zero (the solution of the reduced problem) analogously to what we see in Figure 3.1. *Hint.* A similar argument can be found in [**MS12**].

Why does this standard method give us these oscillations? What has gone wrong? In the next section we will reveal the answers to these questions.

3.1. M-matrices, upwinding

A square matrix $A = (A_{ij})$ is said to be an *M-matrix*[1] if $A_{ij} \leq 0$ for all $i \neq j$ and A^{-1} exists with $(A^{-1})_{ij} \geq 0$ for all i, j. Difference schemes that employ M-matrices are common because they are desirable: they are generally stable and are more amenable to analysis.

Exercise 3.3. Let $A = (A_{ij})$ be an M-matrix. Prove that $A_{ii} > 0$ for all i.

Our central difference scheme above fails to satisfy the M-matrix sign condition on the off-diagonal entries since $B_{i,i+1} > 0$ when ε is small relative to h. If $h\|a\|_\infty \leq 2\varepsilon$, then the sign condition is satisfied, and it turns out that the difference method gives an acceptable computed solution, but to enforce this inequality when ε is small is impractical in many problems (especially in partial differential equations, where multiple dimensions are involved) since it can lead to an intolerable number of mesh points.

The second M-matrix requirement—that A^{-1} exists with $(A^{-1})_{ij} \geq 0$ for all i and j—does not seem easy to verify in practice. Fortunately there are more tractable alternatives, as stated in the next two lemmas.

A square matrix $A = (A_{ij})$ is said to be *strictly diagonally dominant* if $A_{ii} > \sum_{j \neq i} |A_{ij}|$ for all i.

Lemma 3.4. *Suppose that the square matrix $A = (A_{ij})$ satisfies $A_{ij} \leq 0$ for all $i \neq j$. Then A^{-1} exists and $(A^{-1})_{ij} \geq 0$ for all i, j if A is strictly diagonally dominant with $A_{ii} > 0$ for all i.*

Proof. See, e.g., [**QV94**, Lemma 2.1.1]. □

[1] *Historical Note.* The "M" in M-matrix refers to Hermann Minkowski, who studied some of their properties. M-matrices have been exhaustively analysed in the research literature.

Consider a vector $\mathbf{w} = (w_1, w_2, \ldots, w_n) \in \mathbb{R}^n$. By $\mathbf{w} > \mathbf{0}$ we mean that $w_i > 0$ for $i = 1, 2, \ldots, n$. Similarly, $\mathbf{w} \geq \mathbf{0}$ means $w_i \geq 0$ for all i. We set $|\mathbf{w}| = (|w_1|, |w_2|, \ldots, |w_n|)$. The discrete L^∞ norm $\|\cdot\|_{\infty,d}$ for vectors in $\mathbb{R}^n$ is defined by $\|\mathbf{w}\|_{\infty,d} = \max_i |w_i|$. The matrix norm $\|\cdot\|_{\infty,d}$ is the norm induced by the corresponding vector norm $\|\cdot\|_{\infty,d}$; for the $n \times n$ matrix $A = (A_{ij})$ it is the "maximum row sum" norm, viz., $\|A\|_{\infty,d} = \max_i \sum_j |A_{ij}|$. Matrix norms induced by vector norms are discussed in many basic numerical analysis books.

Lemma 3.5. *Suppose that the $n \times n$ matrix $A = (A_{ij})$ satisfies $A_{ij} \leq 0$ for all $i \neq j$. Then A^{-1} exists and $(A^{-1})_{ij} \geq 0$ for all i, j if and only if there exists a vector $\mathbf{w} > 0$ in $\mathbb{R}^n$ such that $A\mathbf{w} > 0$. Furthermore, we have*

$$\|A^{-1}\|_{\infty,d} \leq \frac{\|\mathbf{w}\|_{\infty,d}}{\min_k (A\mathbf{w})_k}. \tag{3.3}$$

Proof. See [**Boh81**] or [**AK90**]. □

One can often construct a vector $\mathbf{w}$ that satisfies the conditions of Lemma 3.5 by first finding a function $w(x)$ such that $w > 0$ and $Lw > 0$, then restricting w to the mesh to form $\mathbf{w}$.

For M-matrices we have the following discrete analogues of Lemma 1.8 and Corollary 1.12.

Lemma 3.6 (Discrete maximum principle)**.** *Let A be an M-matrix. If $\mathbf{w}$ is a vector with $A\mathbf{w} \geq \mathbf{0}$, then $\mathbf{w} \geq \mathbf{0}$.*

Proof. $\mathbf{w} = (A^{-1})(A\mathbf{w}) \geq \mathbf{0}$, because $A^{-1} \geq 0$ and $A\mathbf{w} \geq \mathbf{0}$. □

Lemma 3.7 (Discrete barrier function)**.** *Let A be an M-matrix. If $\mathbf{w}, \mathbf{z}$ are vectors such that $|A\mathbf{w}| \leq A\mathbf{z}$, then $|\mathbf{w}| \leq \mathbf{z}$.*

Proof. Now $A(\mathbf{z} - \mathbf{w}) \geq \mathbf{0}$, so $\mathbf{w} - \mathbf{z} \geq \mathbf{0}$ by Lemma 3.6. Similarly, one has $\mathbf{z} + \mathbf{w} \geq \mathbf{0}$, and the result follows. □

When we take A to be the matrix arising from a discretisation of a boundary value problem, at first sight the boundary data requirement of Corollary 1.12 (the continuous analogue of Lemma 3.7) seems to be missing from Lemma 3.7, but this is deceptive. The first and last rows of A will include this information—see the construction of our matrix B above.

Returning to our difference scheme and its failure to generate an M-matrix, we see that the "incorrect" sign of $B_{i,i+1}$ comes from the central difference approximation $u'(x_i) \approx (u_{i+1}^N - u_{i-1}^N)/(2h)$. This approximation is generally recommended in basic courses in numerical methods because it gives an $\mathcal{O}\left(h^2\right)$ consistency error, but this consistency property is useless when the method is (as we saw) unstable. To cure the instability, for convection-diffusion problems one can approximate $u'(x_i)$ by the *simple upwinding* formula $(u_i^N - u_{i-1}^N)/h$. Although the consistency error is now only $\mathcal{O}(h)$, the ith row of the scheme is

$$\begin{pmatrix} 0\cdots 0 & -\dfrac{\varepsilon}{h^2} - \dfrac{a_i}{h} & \dfrac{2\varepsilon}{h^2} + \dfrac{a_i}{h} + b_i & -\dfrac{\varepsilon}{h^2} & 0\cdots 0 \end{pmatrix},$$

which has the correct sign pattern. Hence, writing B for the associated $(N+1)\times(N+1)$ matrix that incorporates the boundary conditions, one has $B_{ij} \le 0$ for $i \ne j$, as desired.

Lemma 3.8. *The coefficient matrix B for the simple upwind scheme is an M-matrix, and the scheme is uniformly stable with respect to the perturbation parameter*

$$\|u_h\|_{\infty,d} \le C\|Bu_h\|_{\infty,d},$$

with a stability constant C that is independent of ε and h.

Proof. Clearly, $B_{ij} \le 0$ for $i \ne j$. We construct a suitable majorizing vector. Choose $w(x) := 1 + x$, so $Lw(x) \ge \alpha$. Let $\mathbf{v}$ be the restriction of w to the mesh. A quick computation yields $B\mathbf{v} \ge \min\{1, \alpha\}\mathbf{1}$, where $\mathbf{1} = (1, 1, \ldots, 1)^T$. Thus by Lemma 3.5 the matrix B is an M-matrix, and one gets the desired stability bound with stability constant $C = 1/\min\{1, \alpha\}$. □

Simple upwinding for (2.14) uses the one-sided difference $(u_i^N - u_{i-1}^N)/h$ to approximate $u'(x_i)$, but the alternative one-sided difference $(u_{i+1}^N - u_i^N)/h$ would not give the correct sign pattern in the matrix. Upwinding (of which there are many variants) means taking a nonsymmetric finite difference approximation that is *weighted away from the layer*. With simple upwinding, for $\varepsilon \ll h^2$ the scheme almost decouples the boundary condition at $x = 1$ from the values at the interior nodes. This is exactly what is needed to avoid the anomaly described in Remark 3.1.

Remark 3.9. In its various forms, upwinding uses discretisations of the convection term that are suitable for solving the reduced problem (2.18). This is more evident when dealing with the reduced problem (4.3) for convection-diffusion problems posed on domains in two dimensions. The construction and analysis of numerical methods for such "first-order hyperbolic" equations and their nonlinear generalisations has been the subject of much research.

Figure 3.2, where $N = 10$ so the mesh points are $0, 0.1, 0.2, \ldots, 1$, illustrates the difference between the central difference and upwind approximations of $u'(x_{N-1})$ in the typical case when $N \ll 1/\varepsilon$. Clearly, the central difference approximation (the slope of the dashed line through the

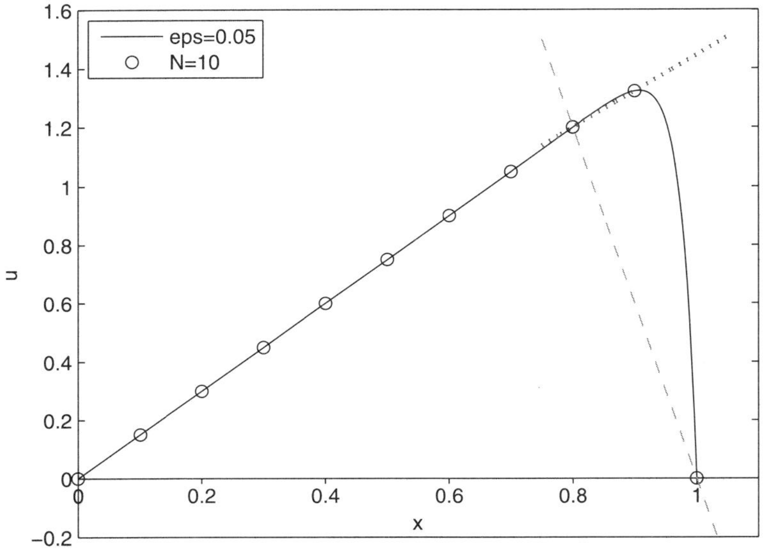

Figure 3.2. Central difference and upwind approximations of $u'(x_{N-1})$

points $(0.8, u(0.8))$ and $(1, 0))$ of $u'(x_{N-1}) = u'(0.9)$ is poor; the simple upwind approximation (the slope of the dotted line through $(0.8, u(0.8))$ and $(0.9, u(0.9)))$ is much better.

Remark 3.10. At first sight it is surprising that, although the singularly perturbed nature of (3.1) comes from the small coefficient ε of the *diffusion* term, the instability difficulties in discretisation arise from how the *convection* term is treated. The reason is that in the classical case $\varepsilon = 1$ the diffusion term stabilizes the numerical method, but this ability is greatly diminished when ε is small, so we must turn to the convection term for assistance in stabilization. In section 3.2 we shall see that convective discretisations such as simple upwinding are in fact equivalent (in a certain sense) to artificially increasing the value of ε.

The first satisfactory investigation into the accuracy of simple upwinding is due to Kellogg and Tsan [**KT78**]. Their delicate analysis derived a tight bound on the consistency error of the method, then converted this to the following convergence result by means of discrete barrier functions. The proof of their result that we now give follows [**KT78**] for the most part; see also [**RST08**, Section I.2.1.2].

Theorem 3.11 (Error bound for simple upwinding on an equidistant mesh). *Let* $\{u_i^N\}_{i=0}^N$ *be the solution to* (3.1) *computed using simple upwinding on an equidistant mesh of diameter* h *with* N *subintervals. Then the error at the* *inner grid points* $\{x_i : i = 1, \ldots, N-1\}$ *satisfies*

$$|u(x_i) - u_i^N| \le \begin{cases} Ch\left[1 + \varepsilon^{-1}\exp\left(-\frac{\beta(1-x_i)}{\varepsilon}\right)\right] & \text{if } h \le \varepsilon, \\ C\left[h + (1 + \frac{\alpha h}{\varepsilon})^{-(N-i)}\right] & \text{if } h > \varepsilon, \end{cases}$$

where $\beta := \ln(1+\alpha)$.

Proof. Let L^N denote the discrete simple upwinding operator. Its consistency error τ_i at each interior mesh point x_i is estimated using Taylor's formula or Peano's theorem, and one obtains

$$(3.4) \qquad |\tau_i| := |L^N u(x_i) - f(x_i)| \le C \int_{x_{i-1}}^{x_{i+1}} \left(\varepsilon |u^{(3)}(t)| + |u^{(2)}(t)|\right) dt.$$

Invoking Theorem 2.27 to bound the terms in (3.4) yields

$$\begin{aligned} |\tau_i| &\le Ch + C\varepsilon^{-2} \int_{x_{i-1}}^{x_{i+1}} \exp\left[-\alpha(1-t)/\varepsilon\right] dt \\ (3.5) \qquad &= Ch + C\varepsilon^{-1} \sinh\left(\frac{\alpha h}{\varepsilon}\right) \exp\left(-\frac{\alpha(1-x_i)}{\varepsilon}\right). \end{aligned}$$

Case $h \le \varepsilon$. Then $\alpha h/\varepsilon \le \alpha$. Now $0 < \sinh t \le Ct$ for $0 < t \le C$, so (3.5) yields

$$(3.6) \qquad |\tau_i| \le Ch\left[1 + \varepsilon^{-2}\exp\left(-\frac{\alpha(1-x_i)}{\varepsilon}\right)\right].$$

We want to apply Lemma 3.7 by constructing a discrete barrier function that will improve the factor ε^{-2} in this bound to ε^{-1}. Now

$$\begin{aligned} L^N\left(1 + \frac{\alpha h}{\varepsilon}\right)^i &\ge \frac{1}{h} \cdot \frac{\alpha h}{\varepsilon}(a_i - \alpha)\left(1 + \frac{\alpha h}{\varepsilon}\right)^{i-1} \\ &\ge \frac{C}{\varepsilon}\left(1 + \frac{\alpha h}{\varepsilon}\right)^{i-1} \\ &\ge \frac{C}{\varepsilon}\left(1 + \frac{\alpha h}{\varepsilon}\right)^{i} \end{aligned}$$

for some constant $C > 0$, using $h \le \varepsilon$. Thus

$$\begin{aligned} L^N\left(1 + \frac{\alpha h}{\varepsilon}\right)^{-(N-i)} &\ge \frac{C}{\varepsilon}\left(1 + \frac{\alpha h}{\varepsilon}\right)^{-(N-i)} \\ &\ge \frac{C}{\varepsilon}\left(e^{\alpha h/\varepsilon}\right)^{-(N-i)} \\ &= \frac{C}{\varepsilon}\exp\left(-\frac{\alpha(1-x_i)}{\varepsilon}\right). \end{aligned}$$

But (3.6) now implies that one can choose a constant $C^* > 0$ such that

$$w_i^* := C^* h \left[x_i + \varepsilon^{-1} \left(1 + \frac{\alpha h}{\varepsilon}\right)^{-(N-i)} \right]$$

is a barrier function for τ_i. Hence

$$|u_i - u_i^N| \le w_i^* \le C^* h \left[1 + \varepsilon^{-1} \left(1 + \frac{\alpha h}{\varepsilon}\right)^{-(N-i)} \right] \text{ for all } i. \tag{3.7}$$

Set $\beta = \ln(1+\alpha)$ and $\phi(t) = e^{\beta t} - (1+\alpha t)$ for $0 \le t \le 1$. Then $\phi(0) = \phi(1) = 0$ and $\phi''(t) = \beta^2 e^{\beta t} > 0$ for $0 < t < 1$. Consequently, $\phi(t) \le 0$ for $0 \le t \le 1$, i.e., $e^{\beta t} \le 1 + \alpha t$ for $0 \le t \le 1$. Thus (3.7) implies that

$$|\tau_i| \le C^* h \left[1 + \varepsilon^{-1} \left(e^{\beta h/\varepsilon}\right)^{-(N-i)} \right] = C^* h \left[1 + \varepsilon^{-1} \exp\left(-\frac{\beta(1-x_i)}{\varepsilon} \right) \right].$$

Case $h > \varepsilon$. This case (which is the case one meets in practice) is more difficult. We begin with a simple decomposition of u. Set

$$v(x) = \frac{\varepsilon u'(1)}{a(1)} \exp\left(-\frac{a(1)(1-x)}{\varepsilon} \right) \quad \text{and } z(x) = u(x) - v(x) \text{ for } x \in [0,1].$$

Thus $u = v + z$. Note that $|\varepsilon u'(1)| \le C$ by Theorem 2.27. Now v is evidently a layer function, and

$$z'(1) = u'(1) - v'(1) = 0$$

by the definition of v, so we expect z to be better behaved than u. (One could instead use one of the decompositions of section 2.4; the above simpler decomposition comes from [**KT78**].) The bound $|z(0)| \le C$ follows from Lemma 2.17. Also,

$$\begin{aligned}
&|Lz(x)| \\
&= \left| f(x) - \frac{\varepsilon u'(1)}{a(1)} \left[-\varepsilon \frac{(a(1))^2}{\varepsilon^2} + \frac{a(x)a(1)}{\varepsilon} + b(x) \right] \exp\left(-\frac{a(1)(1-x)}{\varepsilon} \right) \right| \\
&\le C[1 + \varepsilon^{-1} e^{-a(1)(1-x)/\varepsilon}] \\
&\le C[1 + \varepsilon^{-1} e^{-\alpha(1-x)/\varepsilon}].
\end{aligned}$$

That is, z satisfies the hypotheses of Remark 2.33 and Exercise 2.35, whence

$$|z^{(j)}(x)| \le C[1 + \varepsilon^{1-j} e^{-\alpha(1-x)/\varepsilon}] \quad \text{for } x \in [0,1]. \tag{3.8}$$

Now define discrete functions $\{v_i^N\}_{i=0}^N$ and $\{z_i^N\}_{i=0}^N$ by

$$L^N v_i^N = Lv(x_i) \text{ and } L^N z_i^N = Lz(x_i) \quad \text{for } i = 1, \dots, N-1,$$

with $v_0^N = v(x_0)$, $v_N^N = v(x_N)$, $z_0^N = z(x_0)$, $z_N^N = z(x_N)$. Then for each i one has

$$|u(x_i)-u_i^N| = |v(x_i)+z(x_i)-(v_i^N+z_i^N)| \le |v(x_i)-v_i^N|+|z(x_i)-z_i^N|. \tag{3.9}$$

For the consistency error $\tau_i(z)$ associated with z, like the derivation of (3.5) one gets

$$|\tau_i(z)| \le Ch + C\sinh\left(\frac{\alpha h}{\varepsilon}\right)\exp\left(-\frac{\alpha(1-x_i)}{\varepsilon}\right).$$

(Unlike (3.5) there is no multiplying factor ε^{-1} here because of the extra positive power of ε in (3.8).) As $h > \varepsilon$, we use the inequality $\sinh t \le 2e^t$ for $t > 0$. Hence

$$\begin{aligned} |\tau_i(z)| &\le Ch + C\exp\left(-\frac{\alpha(1-x_{i+1})}{\varepsilon}\right) \\ &= Ch + C\left(e^{-\alpha h/\varepsilon}\right)^{N-(i+1)} \\ &\le Ch + C\left(1+\frac{\alpha h}{\varepsilon}\right)^{-[N-(i+1)]} \end{aligned} \tag{3.10}$$

by an elementary inequality. But a computation shows that

$$L^N\left(1+\frac{\alpha h}{\varepsilon}\right)^i \ge \frac{1}{h}\cdot\frac{\alpha h}{\varepsilon}(a_i-\alpha)\left(1+\frac{\alpha h}{\varepsilon}\right)^{i-1} \ge \frac{C}{h}\left(1+\frac{\alpha h}{\varepsilon}\right)^i, \tag{3.11}$$

using $h > \varepsilon$. It follows from (3.10) and (3.11) that one can choose a constant C such that

$$Ch\left[x_i + \left(1+\frac{\alpha h}{\varepsilon}\right)^{-[N-(i+1)]}\right]$$

is a barrier function for $z(x_i) - z_i^N$. Thus

$$|z(x_i) - z_i^N| \le Ch\left[x_i + \left(1+\frac{\alpha h}{\varepsilon}\right)^{-[N-(i+1)]}\right] \le Ch. \tag{3.12}$$

Now we deal with v. Observe first that for all i,

$$v(x_i) \le C\exp\left(-\frac{\alpha(1-x_i)}{\varepsilon}\right) = \left(e^{-\alpha h/\varepsilon}\right)^{N-i} \le C\left(1+\frac{\alpha h}{\varepsilon}\right)^{-(N-i)}.$$

From the definition of v, one can show readily that $|Lv(x)| \le C\varepsilon^{-1}|v(x)|$. Hence

$$|L^N v_i^N| = |Lv(x_i)| \le \frac{C}{\varepsilon}v(x_i) \le \frac{C}{\varepsilon}\left(e^{-\alpha h/\varepsilon}\right)^{N-i} \le \frac{C}{h}\left(1+\frac{\alpha h}{\varepsilon}\right)^{-(N-i)}$$

by Exercise 3.13. Appealing again to (3.11) and the discrete comparison principle, we obtain

$$|v_i^N| \le C\left(1+\frac{\alpha h}{\varepsilon}\right)^{-(N-i)}.$$

Thus

$$|v(x_i) - v_i^N| \le |v(x_i)| + |v_i^N| \le C\left(1+\frac{\alpha h}{\varepsilon}\right)^{-(N-i)}. \tag{3.13}$$

Combining the inequalities (3.9), (3.12), and (3.13) completes the proof[2] for the case $h > \varepsilon$. □

Remark 3.12. The discrete barrier function $(1+\alpha h/\varepsilon)^{-(N-i)}$ that is used twice in the proof of Theorem 3.11 is a solution of $L^N\phi_i = 0$ when $a \equiv \alpha$, $b \equiv 0$. It is an approximation of the continuous layer function Φ that is a solution of $L\Phi = 0$.

Exercise 3.13. Here is the proof of the inequality

$$\varepsilon^{-1}\left(e^{-\alpha h/\varepsilon}\right)^{N-i} \le Ch^{-1}\left(1+\frac{\alpha h}{\varepsilon}\right)^{-(N-i)} \tag{3.14}$$

for $h > \varepsilon$ from [**KT78**, p. 1031]. Can you give a shorter proof?

Since $e^t \ge 1+t$ and $t(1+t)e^{-t} \le C$ for all $t \ge 0$ where C is some constant, it follows that $t \ge \ln(1+t)$ and $\ln t \le t - \ln(1+t) + \ln C$. Hence

$$\ln t \le \frac{1-x_i}{h}\left[t - \ln(1+t)\right] + \ln C.$$

Let $t = \alpha h/\varepsilon$, then

$$\ln\frac{\alpha h}{\varepsilon} - \frac{1-x_i}{h}\cdot\frac{\alpha h}{\varepsilon} \le -\frac{1-x_i}{h}\ln\left(1+\frac{\alpha h}{\varepsilon}\right) + \ln c.$$

Taking the exponential of both sides, we get (3.14).

Exercise 3.14. Assume $h > \varepsilon$. Prove the inequality

$$\left(1+\frac{\alpha h}{\varepsilon}\right)^{-(N-i)} \le \exp\left(-\frac{\alpha(1-x_i)}{\alpha h+\varepsilon}\right),$$

where $i \in \{1, 2, \ldots, N-1\}$. *Hint.* First prove the inequality $\ln(1-t) < -t$ for $0 < t < 1$.

[2] *Historical Note.* Did you think that the proof of Theorem 3.11 was complicated? When it appeared in 1978 in [**KT78**], it was greeted with relief and joy by numerical analysts as a significant simplification of the then-standard method of analysis of finite difference methods for convection-diffusion problems: a horrendously complicated approach known as the double-mesh principle. By their elegant and powerful use of discrete barrier functions, Kellogg and Tsan [**KT78**] revolutionised numerical analysis in this area. Up to the present day, almost every finite difference analysis of a convection-diffusion two-point boundary value problem uses discrete barrier functions.

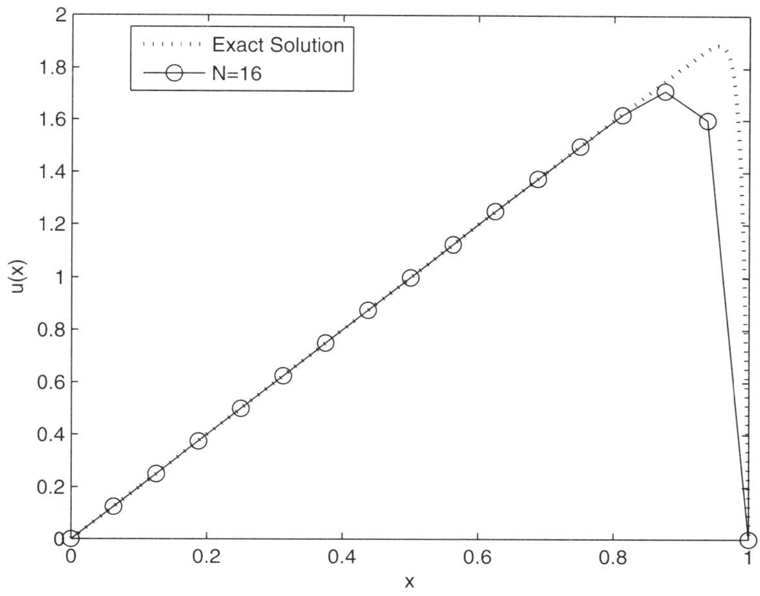

Figure 3.3. Example 1.1 with $\varepsilon = 0.01$; solution computed using simple upwinding with $N = 16$

By invoking Exercise 3.14, the bound of Theorem 3.11 in the case $h > \varepsilon$ can be replaced by

$$C \exp\left(- \frac{\alpha(1 - x_i)}{\alpha h + \varepsilon} \right).$$

If $h \gg \varepsilon$, then Theorem 3.11 implies that the upwind scheme yields an accurate solution at all points. But when $h \approx \varepsilon$, then the scheme is only $\mathcal{O}(1)$-accurate at interior mesh points that lie close to or inside the boundary layer; see Figure 3.3.

Remark 3.15. This type of error behaviour can lead to disconcerting and puzzling results in numerical experiments with simple upwinding (and other forms of upwinding). Suppose that for a given convection-diffusion problem, initially one has an equidistant mesh with $h \gg \varepsilon$, so all mesh points in (0,1) lie well outside the layer. Now consider what happens if we repeatedly bisect each interval and compute a fresh solution. At first the interior mesh points remain outside the layer, so by Theorem 3.11 the numerical results show that the maximum nodal error is small. But as we continue to bisect the mesh, eventually mesh points begin to move into the layer—where the accuracy of the computed solution is only $\mathcal{O}(1)$—so at this stage mesh bisection causes the maximum nodal error to *increase!* See Figure 3.4 in which the maximum nodal error (i.e., the error measured in $\|\cdot\|_{\infty,d}$) for $N = 16$ is *greater* than the error for $N = 8$.

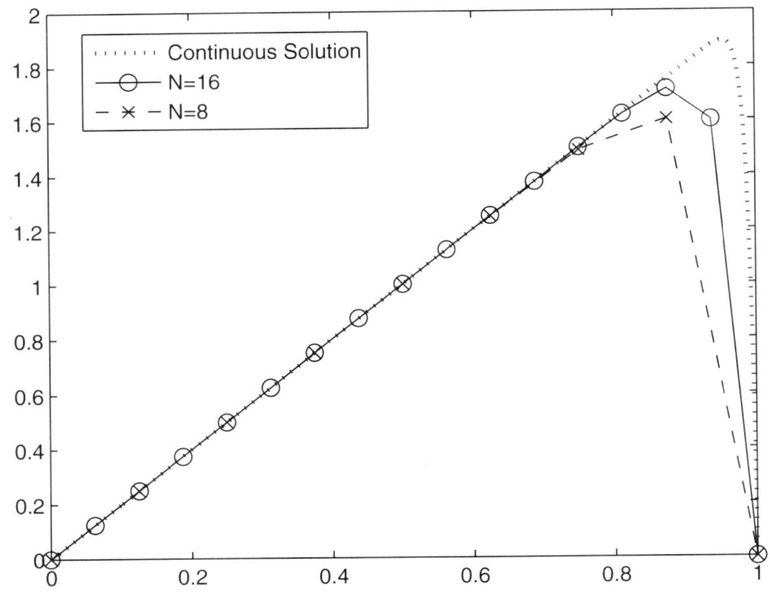

Figure 3.4. Example 1.1 with $\varepsilon = 0.01$; solution computed by simple upwinding for $N = 8, 16$

Exercise 3.16. Suppose that the Dirichlet boundary condition $u(1) = 0$ in (3.1) is changed to the Neumann condition $u'(1) = k$ for some constant k. Then Remark 2.33 gives bounds on the derivatives of u showing that the layer at $x = 1$ is now weaker. Suppose we solve this problem using simple upwinding on an equidistant mesh $x_0, \ldots, x_N$ of diameter h, approximating the Neumann condition by $(u_N^N - u_{N-1}^N)/h = k$. Modify the arguments of Theorem 3.11 to show that $\max_i |u(x_i) - u_i^N| \le Ch$ for some constant C.

3.2. Artificial diffusion

While upwinding does remove unnatural oscillations from the computed solution, one pays a price for this: the layers in the computed solution are excessively smeared, i.e., they are not as steep as they should be; see Figure 3.3. To put this another way, upwinding seems to produce an accurate solution for a different problem where the diffusion coefficient is much greater than ε. We now make this visual observation more precise.

The simple upwinding discretization of $(-\varepsilon u'' + au' + bu)(x_i)$ is

$$\begin{aligned}
&\frac{-\varepsilon}{h^2}(u_{i+1}^N - 2u_i^N + u_{i-1}^N) + \frac{a_i}{h}(u_i^N - u_{i-1}^N) + b_i u_i^N \\
&\quad = -\left(\varepsilon + \frac{ha_i}{2}\right)\frac{1}{h^2}(u_{i+1}^N - 2u_i^N + u_{i-1}^N) + \frac{a_i}{2h}(u_{i+1}^N - u_{i-1}^N) + b_i u_i^N .
\end{aligned}$$

That is, upwinding applied to the original differential equation $Lu = f$ is *exactly the same method* as standard central differencing applied to the modified differential equation $\tilde{L}u := -(\varepsilon + ha/2)u'' + au' + bu = f$.

The diffusion coefficient in this modified differential equation is so large (relative to ε) that central differencing produces an M-matrix and yields an approximation of the true solution of $\tilde{L}u = f$, but of course near $x = 1$ this solution is not close to the solution of $Lu = f$.

The amount $ha(x)/2$ by which the diffusion coefficient was apparently increased by upwinding is called the *artificial diffusion* introduced by upwinding.

This relationship between simple upwinding, $Lu = f$ and $\tilde{L}u = f$ opens the door to a flood of possibilities: one can choose a certain amount of artificial diffusion to add to the problem $Lu = f$, then apply a standard (i.e., not designed for convection-diffusion) numerical method, with the aim of retaining stability (i.e., excluding oscillations) while minimizing the smearing of layers in the computed solution. Pursuing this approach turns out to be quite fruitful; in fact, stable numerical methods on uniform meshes for convection-diffusion ordinary differential equations are usually equivalent to modifying the diffusion in the original differential equation then applying a standard method such as central differencing—but for partial differential equations, the connection may be less straightforward.

Exercise 3.17. Samarskiĭ's difference scheme (see [**KT78**]) for (3.1) is

$$\frac{-\varepsilon}{h^2[1+r_i]}(u_{i+1}^N - 2u_i^N + u_{i-1}^N) + \frac{a_i}{h}(u_i^N - u_{i-1}^N) + b_i u_i^N = f_i^N,$$

where $r_i := ha_i/(2\varepsilon)$. Here simple upwinding is used for the convection term. Because, as we now know, this discretisation of the convection term adds too much artificial diffusion, Samarskiĭ's method counters this excess by decreasing the diffusion coefficient in the differential equation.

Show that Samarskiĭ's scheme can be generated by adding $\varepsilon r_i^2/(1 + r_i)$ artificial diffusion to (3.1) then applying central differencing.

If $ha_i \gg \varepsilon$, then r_i is large, and the added artificial diffusion is approximately $\varepsilon r_i = ha_i/2$, so Samarskiĭ's scheme is close to simple upwinding, while if $ha_i \ll \varepsilon$, then $r_i \approx 0$, so the added artificial diffusion is approximately zero and the scheme resembles standard central differencing. Thus Samarskiĭ's scheme is a form of upwinding that interpolates between these two extreme cases. This construction is reasonable; see Example 3.22. A precise convergence result for the scheme is proved in [**KT78**, Theorem 4.2].

To summarize what we have learned about artificial diffusion: a numerical method specially designed for a convection-diffusion problem is usually equivalent to modifying the problem by adding artificial diffusion then applying a standard numerical method. If too little artificial diffusion is added, then the computed solution is often oscillatory, while if too much diffusion is added, then the computed layers are smeared.

Can one add just the right amount of artificial diffusion to (3.1) so that, when central differencing is then applied, one obtains a stable computed solution that does not smear the boundary layer? We will answer this question in the next section.

3.3. Uniformly convergent schemes

We now consider difference schemes on an equidistant mesh that are accurate both outside and inside the boundary layer. A difference scheme on an arbitrary mesh of $N+1$ points is said to be *robust* or *uniformly convergent (with respect to* ε*) of order* $\beta > 0$ *in the discrete* L^∞ *norm* if there exist constants ε_0 and N_0, which are independent of each other, such that the solution $\{u_i^N\}$ of the scheme satisfies

$$|u_i - u_i^N| \le CN^{-\beta} \qquad \text{for } 0 < \varepsilon \le \varepsilon_0, N \ge N_0 \text{ and } i = 0, \dots, N.$$

Here β is some positive constant that is independent both of the mesh and of ε. We remind the reader that a constant denoted by C is also independent both of ε and the mesh.

A uniformly convergent scheme (on an equidistant mesh) must address explicitly the exponential nature of the layer part of the solution u, as the next result shows.

Theorem 3.18 (Two necessary conditions for uniform convergence on an equidistant mesh). *Assume that we have an equidistant mesh of diameter* $h = 1/N$ *for some positive integer* N*. Suppose that a difference scheme for the problem* $-\varepsilon u'' + au' = f$, $u(0) = g_0$, $u(1) = g_1$, *where* a *is a positive constant, can be written in the form*

$$\theta_- u_{i-1}^N + \theta_0 u_i^N + \theta_+ u_{i+1}^N = hf_i \qquad \textit{for } i = 1, \dots, N-1, \tag{3.15a}$$

$$u_0^N = g_0, \; u_N^N = g_1, \tag{3.15b}$$

where each $\theta = \theta(h, \varepsilon)$ *depends only on the ratio* h/ε*. If the scheme is uniformly convergent for some* $\beta > 0$*, then one must have*

$$\theta_- + \theta_0 + \theta_+ = 0 \quad \textit{and} \quad e^{-ah/\varepsilon}\theta_- + \theta_0 + e^{ah/\varepsilon}\theta_+ = 0. \tag{3.16}$$

Proof. The idea of the proof is to use uniform convergence to replace the u_j^N in (3.15) by u_j, then investigate what happens as $h \to 0$ while holding $\varepsilon = h$, so each θ remains constant.

The hypotheses of the theorem imply that in particular the scheme is uniformly convergent for the problem

$$-\varepsilon u'' + au' = 0,\ u(0) = 1,\ u(1) = 0, \tag{3.17}$$

where a is a positive constant, whose solution is

$$u(x) = 1 - \frac{e^{-a(1-x)/\varepsilon} - e^{-a/\varepsilon}}{1 - e^{-a/\varepsilon}} \quad \text{for } x \in [0,1]. \tag{3.18}$$

Let j be a fixed positive integer. Set $i = N - j$, so $N - i$ is fixed. Then taking a limit in (3.15) applied to (3.17), we get

$$0 = \lim_{\varepsilon = h \to 0} \left[\theta_- u_{i-1}^N + \theta_0 u_i^N + \theta_+ u_{i+1}^N\right] = \lim_{\varepsilon = h \to 0} \left[\theta_- u_{i-1} + \theta_0 u_i + \theta_+ u_{i+1}\right]$$

because $|u_i - u_i^N| \le Ch$ for all i and the θ coefficients do not change in value as we take the limit. Hence, substituting from (3.18), one has

$$\begin{aligned}
0 = \lim_{\varepsilon = h \to 0} \Bigg\{ &(\theta_- + \theta_0 + \theta_+) \\
&- \frac{1}{1 - e^{-a/\varepsilon}} \left[\theta_- e^{-a(1-x_{i-1})/\varepsilon} + \theta_0 e^{-a(1-x_i)/\varepsilon} + \theta_+ e^{-a(1-x_{i+1})/\varepsilon}\right] \\
&+ \frac{e^{-a/\varepsilon}}{1 - e^{-a/\varepsilon}} (\theta_- + \theta_0 + \theta_+) \Bigg\} \\
= \theta_- + \theta_0 + \theta_+ &- \lim_{\varepsilon = h \to 0} e^{-a(1-x_i)/\varepsilon} \left[\theta_- e^{-ah/\varepsilon} + \theta_0 + \theta_+ e^{ah/\varepsilon}\right] \\
= \theta_- + \theta_0 + \theta_+ &- e^{-a(1-x_i)/\varepsilon} \left[\theta_- e^{-ah/\varepsilon} + \theta_0 + \theta_+ e^{ah/\varepsilon}\right],
\end{aligned}$$

since $\lim_{\varepsilon = h \to 0} e^{-a/\varepsilon} = 0$ and $e^{-a(1-x_i)/\varepsilon} = e^{-a(N-i)h/\varepsilon}$ does not change in value as we take the limit. But in this equation, $N - i$ is *any* fixed positive integer, so $e^{-a(1-x_i)/\varepsilon}$ can take more than one value. We conclude that $\theta_- + \theta_0 + \theta_+ = 0$ and $\theta_- e^{-ah/\varepsilon} + \theta_0 + \theta_+ e^{ah/\varepsilon} = 0$. □

See [**Sty03a**] for a generalization of Theorem 3.18.

The hypothesis of Theorem 3.18, that each θ in (3.15a) depends only on the ratio h/ε, is not restrictive; experience shows that almost all difference schemes for the problem stated in Theorem 3.18 enjoy this property. The first condition in (3.16) is satisfied by all plausible difference schemes (it says merely that the scheme is uniformly convergent when the true solution is a constant; see the proof of Theorem 3.18). It is the second condition of (3.16) that distinguishes uniformly convergent schemes. For example, simple upwinding, central differencing, and Samarskiĭ's difference scheme (Exercise 3.17) all fail to satisfy it.

Exercise 3.19. Show that simple upwinding, central differencing, and the difference scheme of Samarskiĭ (Exercise 3.17) all satisfy the hypothesis of Theorem 3.18 that each θ in (3.15a) depends only on the ratio h/ε.

Exercise 3.20. Theorem 3.18 gives necessary conditions for uniform convergence for three-point schemes. Obtain an analogous result for n-point finite difference schemes with $n \geq 3$.

Exercise 3.21. Theorem 3.18 gives *necessary* conditions for uniform convergence for three-point schemes. Show by a very short argument that these conditions are not *sufficient* conditions for uniform convergence for three-point schemes.

Example 3.22. On equidistant meshes, the best known uniformly convergent scheme for (3.1) is the *Il′in–Allen–Southwell difference scheme.* Allen and Southwell [**AS55**] proposed it without any analysis of its behaviour, then it was independently rediscovered by Il′in [**Il′69**], who gave a complicated proof of its convergence. The scheme is

$$(3.19)\quad 2-\frac{a_i e^{\rho_i}}{h(e^{\rho_i}-1)}u_{i-1}^N+\left[\frac{a_i(e^{\rho_i}+1)}{h(e^{\rho_i}-1)}+b_i\right]u_i^N-\frac{a_i}{h(e^{\rho_i}-1)}u_{i+1}^N=f_i$$

for $i=1,\ldots,N-1$, where $\rho_i = ha_i/\varepsilon$, with $u_0^N = u_N^N = 0$. It computes $\{u_i\}$ *exactly* in the special case where a and f are constants and $b \equiv 0$. This scheme can be generated in a wide variety of ways [**Roo94**]. In [**KT78**] discrete barrier functions were introduced for the first time in the convection-diffusion literature to show that the solution $\{u_i^N\}$ computed by the scheme is first-order uniformly convergent: $|u_i - u_i^N| \leq CN^{-1}$ for all i.

When $ha_i \gg \varepsilon$, the scheme is close to simple upwinding, while if $ha_i \ll \varepsilon$, the scheme resembles central differencing.

Exercise 3.23. Consider the differential equation $-\varepsilon u'' + au' = f$, with a and f constant and $u(0) = u(1) = 0$. On an equidistant mesh of width $h = 1/N$, suppose that the three-point difference scheme (3.15) is required to compute u *exactly* at each mesh point. Determine the conditions that the coefficients θ_-, θ_0, and θ_+ of (3.15) must satisfy; solve these conditions for θ_-, θ_0, and θ_+, and observe that you get (3.19).

Exercise 3.24. Show that the Il′in–Allen–Southwell scheme can be generated by applying central differencing to the modified differential equation

$$-\varepsilon\left(\frac{ha(x)}{2\varepsilon}\coth\frac{ha(x)}{2\varepsilon}\right)u''(x)+a(x)u'(x)+b(x)u(x)=f(x).$$

Show that $[ha(x)/(2\varepsilon)]\coth[ha(x)/(2\varepsilon)] > 1$. This indicates that artificial diffusion (see section 3.2) has been added to the original problem.

Exercise 3.25. Prove that the coefficient matrix associated with the Il′in–Allen–Southwell scheme is an M-matrix. *Hint.* See the proof of Lemma 3.8.

The more complicated El Mistikawy–Werle three-point scheme has the form

$$r_i^- u_{i-1}^N + r_i^0 u_i^N + r_i^+ u_{i+1}^N = q_i^- f_{i-1} + q_i^0 f_i + q_{i+1}^+ f_{i+1} \quad \text{for } i = 1, \ldots, N-1.$$

It achieves second-order uniform convergence on equidistant meshes, i.e., $\max_i |u_i - u_i^N| \le CN^{-2}$. See [**RST08**, Section I.2.1.3] for more information about this scheme and the Il′in–Allen–Southwell scheme.

Numerical methods like these, whose coefficients involve exponential functions of h/ε, are known collectively as *exponentially fitted* schemes. Exponential fitting is the mainstay of the FEM package PLTMG and is widely used in semiconductor device modelling, where the Il′in–Allen–Southwell scheme is known as the Scharfetter–Gummel scheme. A recent related idea is the *tailored finite point method* [**HH14**], where local solutions of the differential equation are used to generate finite difference schemes.

Exercise 3.26. Find a connection between the Il′in–Allen–Southwell scheme and the tailored finite point method of [**HH14**].

Remark 3.27. For the reaction-diffusion problems considered in Remarks 2.37 and 2.50, the standard three-point discretisation of u'' on an equidistant mesh produces a difference scheme whose matrix has ith row given by (3.2) with $a \equiv 0$. This is easily verified to be an M-matrix using Lemma 3.5 (we leave this as an exercise), so its computed solutions are stable. Nevertheless, the method is not uniformly convergent because one can prove an analogue of Theorem 3.18 showing that only schemes whose coefficients have a certain exponential property can be uniformly convergent. An exponentially fitted uniformly convergent scheme for reaction-diffusion problems is analysed in [**OS86**].

Exercise 3.28. Prove an analogue of Theorem 3.18 for reaction-diffusion problems.

3.4. Shishkin meshes

When solving numerically a convection-diffusion problem, it seems reasonable to cluster mesh points in the layer—where the solution $u(x)$ is most troublesome—instead of spreading them equidistantly over [0,1]. This approach is an alternative to the exponential fitting on equidistant meshes that was discussed in section 3.3.

Graded meshes, where the mesh width gets finer and finer as one moves closer and closer to $x = 1$, have been advocated by several authors; see [**Lin10, Roo12**] and [**RST08**, Section I.2.4.1] for references. A well-known example of this class is the Bakhvalov mesh, which is discussed at some length in [**Lin10**]. But convergence analyses on graded meshes can be very delicate, so we shall concentrate here on a simpler piecewise-equidistant mesh that is originally due to Shishkin and which has been used by many researchers. The books by Shishkin et al. [**FHM+00**, **MOS12**, **SS09**] are devoted entirely to the use and analysis of this mesh. See [**KO10**, **Lin10**, **Roo12**, **RST08**] for references to singularly perturbed problems where the Shishkin mesh has been used.

Consider the convection-diffusion problem (3.1). For a full analysis that is valid for all values of ε and N, set $\sigma = \min\{1/2,\ (2/\alpha)\varepsilon \ln N\}$. In our exposition we shall assume that $\sigma = (2/\alpha)\varepsilon \ln N$, as $\sigma = 1/2$ occurs only when N is exponentially large relative to ε, which is rare in practice. Then the *mesh transition point*—which separates the fine and coarse portions of the Shishkin mesh—is defined to be $1 - \sigma$; typically it lies close to 1. Let N be an even integer. Divide each of $[0, 1 - \sigma]$ and $[1 - \sigma, 1]$ by an equidistant mesh with $N/2$ subintervals; see Figure 3.5.

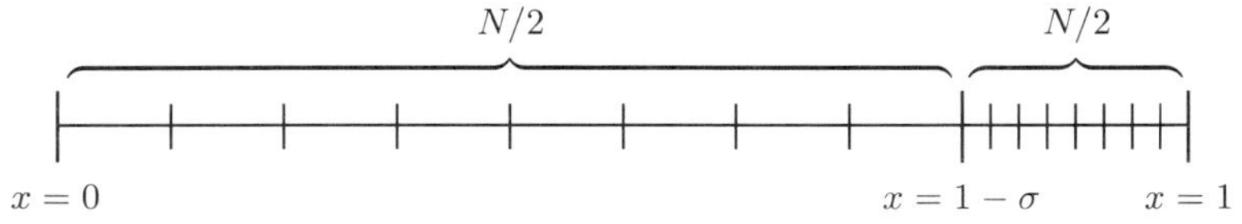

Figure 3.5. Shishkin mesh for convection-diffusion with $N = 16$

The coarse part of this Shishkin mesh has spacing $H := 2(1-\sigma)/N$, so $N^{-1} \le H \le 2N^{-1}$. The fine part has spacing $h := 2\sigma/N = (4/\alpha)\varepsilon N^{-1} \ln N$, so $h \ll \varepsilon$. On the mesh, $x_i = iH$ for $i = 0, \ldots, N/2$ and $x_i = 1 - (N-i)h$ for $i = N/2+1, \ldots, N$. Set $h_i = x_i - x_{i-1}$ for each i.

Remark 3.29. Nonequidistant meshes for convection-diffusion problems are sometimes described as "layer-resolving" meshes. One might presume that this terminology means that wherever the derivatives of $u(x)$ are large, the mesh is fine. But the Shishkin mesh does *not* fully resolve the layer: for $|u'(x)| \approx C[1 + \varepsilon^{-1}\exp(-\alpha(1-x)/\varepsilon)]$ by Theorem 2.27, so

$$|u'(1-\sigma)| \approx C[1 + \varepsilon^{-1}\exp(-2\ln N)] = C[1 + \varepsilon^{-1}N^{-2}],$$

which can be large since typically $\varepsilon \ll N^{-1}$. Thus $|u'(x)|$ is still large on part of the last coarse-mesh interval $[x_{N/2-1}, x_{N/2}]$.

This large derivative is not an error in the design of the mesh! Shishkin's key insight was that one could achieve satisfactory theoretical and numerical results *without resolving all of the layer*. If one sets out to construct a two-stage piecewise-equidistant mesh as we have done, but with the additional requirement that the mesh be fine enough to control the local truncation error wherever $|u'(x)|$ is very large (i.e., one resolves all of the layer), then the number of mesh points required will have to grow like $\ln(1/\varepsilon)$ as ε gets smaller; see [**RST08**, Remark I.2.85]. Shishkin's construction and analysis enables us to work with a fixed number $(N+1)$ of mesh points that does not increase even if ε is very small.

Remark 3.30. The bound $|u'(x)| \le C[1+\varepsilon^{-1}\exp(-\alpha(1-x)/\varepsilon)]$ of Theorem 2.27 implies that $|u'(x)| \le C$ for $0 \le x \le 1-(1/\alpha)\varepsilon\ln(1/\varepsilon)$. This property is often expressed as "the width of the boundary layer is $\mathcal{O}(\varepsilon\ln(1/\varepsilon))$".

Exercise 3.31. For $0 \le x \le 1-\sigma-H$ (i.e., on the coarse mesh region excluding the rightmost interval $[1-\sigma-H, 1-\sigma]$), use Theorem 2.27 to prove that $|u^{(i)}(x)| \le C$ for $i=0,1,2$ and some constant C. How should you choose σ to get the same result for $i=0,1,2,\ldots,q$?

We now analyse simple upwinding on the Shishkin mesh. For each mesh function $\{v_i\}_{i=0}^N$, set $D_-v_i = (v_i - v_{i-1})/h_i$ and

$$\delta^2 v_i = \frac{2}{h_i + h_{i+1}}\left(D_-v_{i+1} - D_-v_i\right);$$

this is a standard discretization of $v''(x_i)$ on a nonequidistant mesh. Our difference scheme is

$$-\varepsilon\delta^2 u_i^N + a_i D_- u_i^N + b_i u_i^N = f_i \quad \text{for } i=1,\ldots,N-1, \tag{3.20a}$$

$$u_0^N = u_N^N = 0. \tag{3.20b}$$

It is straightforward to check (cf. Lemma 3.8) that the matrix L^N associated with (3.20) is an M-matrix. To investigate the convergence of the method, recall the Shishkin decomposition $u = S + E$ of Theorem 2.48 and split the discrete solution $\{u_i^N\}$ in an analogous manner: define $\{S_i^N\}$ and $\{E_i^N\}$ by

$$L^N S_i^N = (LS)_i = f_i \text{ for } i=1,\ldots,N-1, \qquad S_0^N = S(0),\ S_N^N = S(1),$$
$$L^N E_i^N = (LE)_i = 0 \text{ for } i=1,\ldots,N-1, \qquad E_0^N = E(0),\ E_N^N = E(1).$$

Then $u_i^N = S_i^N + E_i^N$ for all i, and

$$|u_i - u_i^N| = |(S+E)_i - (S_i^N + E_i^N)| \le |S_i - S_i^N| + |E_i - E_i^N|. \tag{3.21}$$

We shall bound each right-hand side term separately.

Lemma 3.32. *There exists a constant C_0 such that*

$$|S_i - S_i^N| \le C_0 N^{-1} \quad \textit{for } i=0,\ldots,N.$$

Proof. As the derivatives of S are bounded, a standard consistency error analysis shows that

$$
\begin{aligned}
|L^N(S_i - S_i^N)| &= |L^N S_i - (LS)_i| \\
&\le 2\varepsilon \int_{x_{i-1}}^{x_{i+1}} |S'''(x)|\,dx + a_i \int_{x_{i-1}}^{x_i} |S''(x)|\,dx \qquad (3.22) \\
&\le C(x_{i+1} - x_{i-1}) \le CN^{-1}
\end{aligned}
$$

for $i = 1, \ldots, N-1$. Set $w_i = C_0 N^{-1} x_i$ for all i, where the positive constant C_0 will be chosen so that $\{w_i^N\}$ is a discrete barrier function for $\{S_i - S_i^N\}$. Now

$$L^N w_i = a_i C_0 N^{-1} + b_i w_i > \alpha C_0 N^{-1} \ge |L^N S_i - (LS)_i|$$

by (3.22), provided that C_0 is a sufficiently large constant. Clearly, $w_0 = 0 = |S_0 - S_0^0|$ and $w_N = C_0 N^{-1} \ge 0 = |S_N - S_N^N|$. Thus Lemma 3.7 can be applied, and we get $|S_i - S_i^N| \le w_i \le C_0 N^{-1}$ for all i, as desired. $\square$

To bound $|E_i - E_i^N|$ one appeals again to Lemma 3.7, but the approach cannot be as direct as Lemma 3.32 because $E(x)$ has large derivatives on part of the coarse mesh (see Remark 3.29). Instead, we show first by two separate calculations that $|E_i|$ and $|E_i^N|$ are small on $[0, 1-\sigma]$ because they decay rapidly away from $x = 1$, so $|E_i - E_i^N|$ is small on $[0, 1-\sigma]$. In particular this implies that $|E_i - E_i^N|$ is small when $i = N/2$ (i.e., at the transition point $1-\sigma$). Then on $[1-\sigma, 1]$ the mesh is so fine that it compensates for the large derivatives of u and, consequently, $|E_i - E_i^N|$ can be bounded by a consistency error analysis like that of Lemma 3.32 using the previous bound on $|E_{N/2} - E_{N/2}^N|$.

Note that each of these "consistency and barrier functions imply convergence" arguments is a manifestation of the "consistency and stability imply convergence" principle that is standard in finite difference analyses.

From (2.33) and the definition of the transition point $1-\sigma$,

$$|E_i| \le Ce^{-\alpha(1-(1-\sigma))/\varepsilon} = CN^{-2} \le CN^{-1} \quad \text{for } i = 0, \ldots, N/2. \qquad (3.23)$$

In the next lemma a discrete barrier function is used to show that $|E_i^N|$ is small when $i \le N/2$, like $|E_i|$. Set

$$Z_i = \prod_{j=1}^{i} \left(1 + \frac{\alpha h_j}{2\varepsilon}\right) \quad \text{for } i = 0, \ldots, N,$$

with the standard convention that when $i = 0$, this product is equal to 1.

Exercise 3.33. Show that $Z_i \le \exp(\alpha x_i/(2\varepsilon))$ for $i = 0, 1, , \ldots, N$.

Lemma 3.34. *There exists a constant C such that*

$$|E_i^N| \le CN^{-1} \quad \textit{for } i = 0, \dots, N/2.$$

Proof. For $i = 1, \dots, N$, a calculation shows that there exists a constant $C_1 > 0$ such that

$$L^N Z_i \ge \frac{C_1}{\max\{\varepsilon, h_i\}} Z_i. \tag{3.24}$$

Now $e^t \ge 1 + t$ for all $t \ge 0$, so

$$\frac{Z_i}{Z_N} = \prod_{j=i+1}^{N} \left(1 + \frac{\alpha h_j}{2\varepsilon}\right)^{-1} \ge \prod_{j=i+1}^{N} e^{-\alpha h_j/(2\varepsilon)} = e^{-\alpha(1-x_i)/(2\varepsilon)}. \tag{3.25}$$

Set $Y_i = C_2 Z_i/Z_N$ for $i = 0, \dots, N$. Then $L^N Y_i = (C_2/Z_N) L^N Z_i \ge 0 = |L^N E_i^N|$ for $i = 1, \dots, N-1$, by (3.24) and the definition of $\{E_i^N\}$. Also $Y_N = C_2 \ge |E(1)| = |E_N^N|$ if the constant C_2 is chosen sufficiently large, by the bound on $|E(x)|$ given by inequality (2.33). Finally, (3.25) implies that

$$Y_0 = \frac{C_2 Z_0}{Z_N} \ge C_2 e^{-\alpha/(2\varepsilon)} \ge C_2 e^{-\alpha/\varepsilon} \ge |E(0)| = |E_0^N|,$$

provided that the constant C_2 is chosen sufficiently large, where we appealed again to (2.33). Thus we can choose C_2 so that the conditions of Lemma 3.7 are satisfied (i.e., $\{Y_i\}$ is a discrete barrier function for $\{E_i^N\}$), and it follows that

$$|E_i^N| \le Y_i = \frac{C_2 Z_i}{Z_N} \quad \text{for all } i. \tag{3.26}$$

But for $i = 0, \dots, N/2$,

$$\frac{Z_i}{Z_N} \le \frac{Z_{N/2}}{Z_N} = \prod_{j=1+N/2}^{N} \left(1 + \frac{\alpha h}{2\varepsilon}\right)^{-1} = \left(1 + 2N^{-1} \ln N\right)^{-N/2}$$
$$\le N^{-1} e^{(\ln N)^2/N} \le CN^{-1}$$

for some constant C (to prove the penultimate inequality, see Exercise 3.36). Combining this inequality with (3.26), the proof is complete. □

Exercise 3.35. Prove inequality (3.24).

Exercise 3.36. Prove the inequality $\left(1 + 2N^{-1} \ln N\right)^{-N/2} \le N^{-1} e^{(\ln N)^2/N}$ by first proving that $\ln(1+t) \ge t - t^2/2$ for $t \ge 0$.

Corollary 3.37. *There exists a constant C such that*

$$|E_i - E_i^N| \le CN^{-1} \quad \textit{for } i = 0, \dots, N/2.$$

Proof. This is immediate from (3.23) and Lemma 3.34. □

It remains only to bound $|E_i - E_i^N|$ for $i > N/2$, i.e., on the fine mesh.

Lemma 3.38. *There exists a constant* C *such that*

$$|E_i - E_i^N| \le CN^{-1}\ln N \quad \textit{for } i = N/2+1,\dots,N.$$

Proof. We shall apply a discrete barrier function argument at the nodes $\{x_i\}_{i=N/2}^N$ by considering the discretization of a two-point boundary value problem on the interval $[1-\sigma, 1]$. Observe that when L^N is restricted to the interior nodes of this interval, it still yields an M-matrix.

Recalling the bounds on $|E^{(j)}(x)|$ in (2.33), a standard consistency error analysis shows that for $i = N/2+1,\dots,N-1$,

$$\begin{aligned}
|L^N(E_i - E_i^N)| &= |L^N E_i - (LE)_i| \\
&\le 2\varepsilon \int_{x_{i-1}}^{x_{i+1}} |E'''(x)|\,dx + a_i \int_{x_{i-1}}^{x_i} |E''(x)|\,dx \\
&\le C \int_{x_{i-1}}^{x_{i+1}} \varepsilon^{-2} e^{-\alpha(1-x)/\varepsilon}\,dx \\
&= C\varepsilon^{-1} e^{-\alpha(1-x_i)/\varepsilon} \sinh(\alpha h/\varepsilon) \\
&\le C\varepsilon^{-1} N^{-1}(\ln N) e^{-\alpha(1-x_i)/\varepsilon},
\end{aligned}$$

since $\sinh(\alpha h/\varepsilon) = \sinh(4N^{-1}\ln N) \le CN^{-1}\ln N$ for all $N \ge 2$.

Set $\phi_i = C_3 N^{-1}(\ln N)(1 + Z_i/Z_N)$ for $i = N/2,\dots,N$, where the constant C_3 will be chosen later. By (3.24) and (3.25),

$$\begin{aligned}
L^N\phi_i &\ge C_3 N^{-1}(\ln N)(L^N Z_i)/Z_N \\
&\ge C_3 C_1 \varepsilon^{-1} N^{-1}(\ln N) Z_i/Z_N \\
&\ge C_3 C_1 \varepsilon^{-1} N^{-1}(\ln N) e^{-\alpha(1-x_i)/(2\varepsilon)}
\end{aligned}$$

for $i = N/2+1,\dots,N$. Consequently, $L^N\phi_i \ge |L^N(E_i - E_i^N)|$ if the constant C_3 is sufficiently large. Furthermore, we can choose C_3 such that

$$\phi_{N/2} = C_3 N^{-1}(\ln N)(1 + Z_{N/2}/Z_N) \ge C_3 N^{-1}(\ln N) \ge |E_{N/2} - E_{N/2}^N|$$

by Corollary 3.37, and $\phi_N = 2C_3 N^{-1}(\ln N) > 0 = |E_N - E_N^N|$.

Thus $\{\phi_i\}$ is a discrete barrier function for $\{E_i - E_i^N\}$, and Lemma 3.7 now implies that for $i = N/2,\dots,N$, we have $|E_i - E_i^N| \le \phi_i \le 2C_3 N^{-1}\ln N$. □

The final convergence result can now be stated.

Theorem 3.39 (Uniform convergence of simple upwinding on a Shishkin mesh). *There exists a constant* C *such that the solution* $\{u_i^N\}$ *of* (3.20) *satisfies*

$$|u_i - u_i^N| \le CN^{-1}\ln N \quad \textit{for } i = 0,\dots,N.$$

Proof. Combine (3.21), Lemma 3.32, Corollary 3.37, and Lemma 3.38. □

Numerical results in [**FHM+00**] show that Theorem 3.39 is sharp.

Roos [**Roo96**] shows that the condition number of the discrete linear system associated with (3.20) is $\mathcal{O}\left(\varepsilon^{-2}N^2(\ln N)^{-2}\right)$, which is uncomfortably large when ε is small, but that an easy preconditioning by diagonal scaling (i.e., approximate equilibration) reduces this condition number to $\mathcal{O}\left(N^2(\ln N)^{-1}\right)$. This approach is pursued further in [**NSV18**], where preconditioning is used not only to improve the condition number of the matrix but also to facilitate a new convergence analysis of the method that is closer in spirit to classical finite difference analyses.

Remark 3.40. The *precise choice of mesh transition point* $1-\sigma$ in the Shishkin mesh is of both theoretical and computational interest. A careful examination of the proof of Theorem 3.39 [**FHM+00**], or of the exact error in a special case [**KO10**], reveals that σ should have the form $(k/\alpha)\varepsilon\phi(N)$, where $\phi(N)\to\infty$ but $N^{-1}\phi(N)\to 0$ as $N\to\infty$, and k is some constant. The simplest choice for $\phi(N)$ is $\ln N$.

The choice $k=2$ used in our definition of σ enters subtly the proof of Lemma 3.34 during the final chain of inequalities that bound Z_i/Z_N. How to choose k in an optimal way is discussed in [**ST98**]. It is shown there, using an argument resembling our proof of Theorem 3.39, that for a variant of simple upwinding one has

$$|u_i-u_i^N|\le C\max\{N^{-k},kN^{-1}\ln N\}\quad\text{for } i=0,\dots,N.$$

The sharpness of this bound is confirmed by numerical experiments. Consequently choosing k larger than 1 (i.e., placing the transition point a little too far from $x=1$) diminishes only slightly the numerical accuracy of the method, but choosing k smaller than 1 (i.e., placing the transition point a little too close to $x=1$) causes a noticeable deterioration in the numerical rate of convergence.

Remark 3.41. In [**AK96**] it is shown that for *central differencing on a Shishin mesh*, the computed solution $\{u_i^N\}$ satisfies $|u_i-u_i^N|\le CN^{-2}(\ln N)^2$ for all i. The proof requires some ingenuity as the associated matrix is not an M-matrix (on the coarse part of the mesh, its sign pattern is incorrect) and the scheme does not satisfy a discrete maximum principle. In [**Len00**] an alternative analysis of this method is presented that neatly side-steps this obstacle by considering only alternate mesh points, since this will yield a nonoscillatory computed solution—even on an equidistant mesh (consider Figure 3.1). But numerical experience [**LS01b**] with analogues of central differencing for two-dimensional problems reveals that it is quite expensive to solve the discrete linear system efficiently, so we shall not pursue this method further.

Exercise 3.42. Consider (3.20) and the Shishkin mesh of Theorem 3.39. Suppose we use central differencing at the mesh points x_i for $i = N/2 + 1, \dots, N-1$ (i.e., where the mesh is fine) and upwinding at x_i for $i = 1, 2, \dots, N/2$. Show that the matrix associated with this *hybrid* difference scheme is an M-matrix. Sharpen certain estimates in the proof of Theorem 3.39 to get the improved error bound $|u_i - u_i^N| \le CN^{-1}$ for $i = 0, \dots, N$. (This exercise is based on [**LS99**].)

Remark 3.43. Error estimates in various norms for numerical methods on Shishkin meshes usually include a multiplicative factor $(\ln N)^\beta$ for some $\beta > 0$. This factor is unimportant relative to the main convergence factor N^{-k} with $k > 0$, but it does increase the magnitude of the actual errors. If one works with certain graded meshes (e.g., Bakhvalov meshes), then the $\ln N$ factor disappears, so these meshes yield a higher rate of convergence but they are more complicated to construct and analyse. See [**KLS08, Lin10, RST08**] for more information about these meshes.

The result of Theorem 3.39 can be extended to more general forms of upwinding and to other nonequidistant layer-adapted meshes that are designed for convection-diffusion problems. For excellent surveys of such generalizations for problems in one and two dimensions, see [**Lin03, Lin10**].

Remark 3.44. A typical solution of the reaction-diffusion problem considered in Remarks 2.37, 2.50, and 3.27 has boundary layers at $x = 0$ and $x = 1$, and Remark 2.50 gives us precise information about the nature of these layers. Hence, the mesh displayed in Figure 3.6, where $\sigma_0 = \sigma_1 = (2/\beta)\sqrt{\varepsilon}\ln N$, is a suitable Shishkin mesh for this problem. As in the convection-diffusion

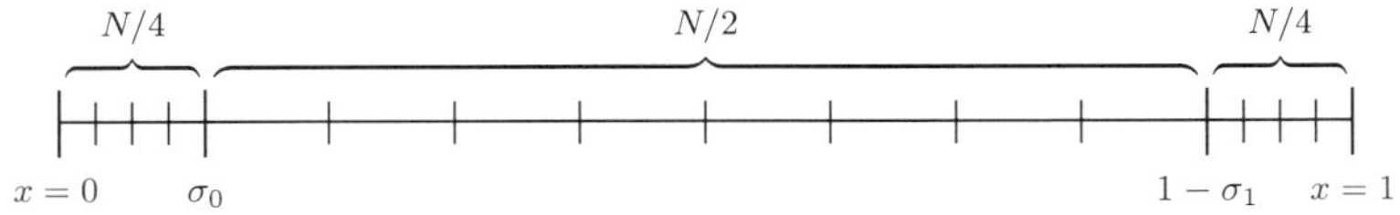

Figure 3.6. Shishkin mesh for reaction-diffusion with $N = 16$

problem, we place half of the N mesh intervals in the coarse mesh; the remaining $N/2$ mesh intervals are divided equally between the two boundary layers.

Analogously to Theorem 3.39, the solution $\{u_i^N\}$ computed by the standard three-point difference scheme obtained on this mesh (set $a \equiv 0$ in (3.20)) satisfies

$$|u_i - u_i^N| \le C(N^{-1}\ln N)^2 \quad \text{for } i = 0, 1, \dots, N;$$

see [**MOS12**, Chapter 6]. Note that this is almost second-order convergence, while Theorem 3.39 gives only almost first-order convergence for the more difficult convection-diffusion problem.

In this chapter we have studied finite difference methods for convection-diffusion problems because they require less background preparation than finite element methods. For two-point boundary value problems, finite differences are just as powerful an approach as finite elements. Perhaps the same is true in two and three dimensions if the domains are rectangular with sides parallel to the coordinate axes. But for problems posed on general two- or three-dimensional domains, finite elements can offer more flexibility; thus in the latter half of the book we shall switch our main focus from finite differences to finite element methods.

Chapter 4

Convection-Diffusion Problems in Two Dimensions

In two dimensions the convection-diffusion equation takes the form

(4.1a) $$Lu(x,y) := -\varepsilon\Delta u(x,y) + \mathbf{a}(x,y)\cdot\nabla u(x,y) + b(x,y)u(x,y) = f(x,y)$$

on $\Omega \subset \mathbb{R}^2$, with

(4.1b) $$u(x,y) = g(x,y) \quad \text{on } \partial\Omega,$$

where $0 < \varepsilon \leq 1$, and the functions $\mathbf{a}, b$, and f are assumed to be Hölder continuous on $\bar{\Omega}$, the closure of Ω. We also assume that $b \geq 0$ on $\bar{\Omega}$. Here Ω is any bounded domain in $\mathbb{R}^2$ with a piecewise Lipschitz-continuous boundary $\partial\Omega$ (e.g., a rectangle or a domain with differentiable boundary). Assume that g is continuous on $\partial\Omega$ except perhaps for a jump discontinuity at a finite number of points. The differential operator L is elliptic, and by Lemma 1.8 it satisfies a maximum principle, so (4.1) has a unique solution in $C^2(\Omega)$; see for example [**GT01**].

4.1. General description

Assume that $\varepsilon \ll 1$ and $|\mathbf{a}| \approx 1$ in (4.1a), so that convection dominates diffusion. In the problems that we consider, the solution $u(x,y)$ of (4.1) has an asymptotic structure similar to that for one-dimensional problems. That is, analogously to the case $k = 0$ in (2.8), one can write $u(x,y)$ as the sum of the solution to a first-order PDE, plus layer(s), plus an $\mathcal{O}(\varepsilon)$ term.

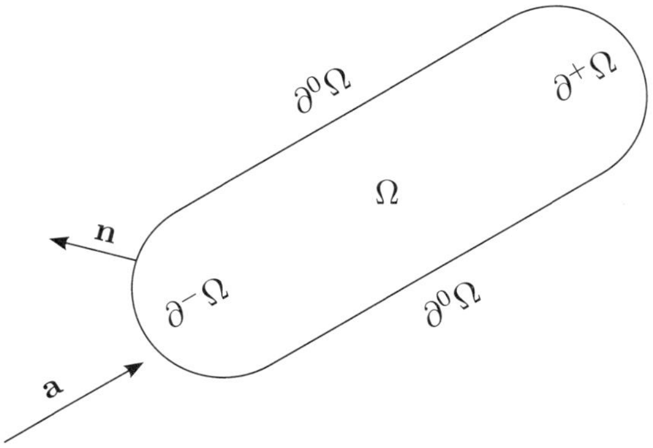

Figure 4.1. Partition of $\partial\Omega$

To make this more precise, divide the boundary $\partial\Omega$ into three parts:

(4.2a) $$\text{inflow boundary } \partial^-\Omega = \{x \in \partial\Omega : \mathbf{a}\cdot\mathbf{n} < 0\},$$

(4.2b) $$\text{outflow boundary } \partial^+\Omega = \{x \in \partial\Omega : \mathbf{a}\cdot\mathbf{n} > 0\},$$

(4.2c) $$\text{characteristic/tangential flow boundary } \partial^0\Omega = \{x \in \partial\Omega : \mathbf{a}\cdot\mathbf{n} = 0\},$$

where $\mathbf{n}$ is the outward-pointing unit normal to $\partial\Omega$; see Figure 4.1.

A typical solution u will have *boundary layers*—narrow regions close to $\partial\Omega$ where $|\nabla u|$ is large—along $\partial^+\Omega$ and $\partial^0\Omega$. As in one-dimensional problems, exceptional Dirichlet boundary conditions g can eliminate these layers; recall Remark 2.2. Also, Neumann boundary conditions on some or all of $\partial^+\Omega$ and $\partial^0\Omega$ mean that layers are no longer visible there (cf. Remark 2.33).

On most of Ω, u is approximately equal to u_0, the solution of the *reduced problem* is

(4.3a) $$\mathbf{a}(x,y)\cdot\nabla u_0(x,y) + b(x,y)u_0(x,y) = f(x,y) \quad \text{on } \Omega,$$

(4.3b) $$u_0 = g \quad \text{on } \partial^-\Omega.$$

This first-order problem is the two-dimensional analogue of (2.18). Write $\mathbf{a} = (a_1, a_2)$. Following the standard theory of first-order PDEs, the *characteristic traces* or *characteristic curves* or *characteristics* of (4.3) are the parameterized curves $(x(t), y(t))$ in Ω defined by

(4.4) $$x'(t) = a_1(x,y), \quad y'(t) = a_2(x,y),$$

with initial data $(x(0), y(0)) = (\hat{x}, \hat{y})$, where $(\hat{x}, \hat{y})$ is any point in $\partial^-\Omega$. One such curve enters Ω from each point in $\partial^-\Omega$. The function $u_0(x,y)$ propagates itself along these curves: on each characteristic, (4.3) simplifies (we leave this as an exercise!) to the ordinary differential equation

$$\frac{du_0(t)}{dt} + bu_0 = f \tag{4.5}$$

with initial data $u_0(0) = g(\hat{x}, \hat{y})$, where we have abused the notation by writing u_0 as a function of t along each characteristic. As in fluid dynamics, the direction of propagation $\mathbf{a}$ is often called the *flow*; this explains the terminology of (4.2).

We shall refer to the characteristics of (4.3) as the *subcharacteristics* of (4.1).

Exercise 4.1. Suppose that

$$-\varepsilon\Delta u + u_x + xu_y = 3 \quad \text{on } \Omega := (0,1) \times (0,1),$$

with $u = 1$ on the inflow boundary $\partial^-\Omega$. Compute the subcharacteristics of this convection-diffusion problem and, hence, show that the solution of the associated reduced problem is

$$u_0(x,y) = \begin{cases} 3x + 1 & \text{if } y \geq x^2/2, \\ 3\left(x - \sqrt{\frac{x^2}{2} - y} + 1\right) & \text{if } y < x^2/2, \end{cases} \qquad \text{for all } (x,y) \in \Omega.$$

Just as in one dimension, boundary layers occur where there is a mismatch between the reduced solution u_0 and the boundary data. This can happen only along $\partial^+\Omega$ and $\partial^0\Omega$. While all layers look much the same when plotted, there can nevertheless be significant analytical differences between them.

Layers along $\partial^+\Omega$ are called *regular* or *exponential boundary layers.* Writing $\vec{n} = (n_1, n_2)$ for the unit outward-pointing normal to $\partial\Omega$, then near $\partial^+\Omega$, exponential layers are essentially multiples of the function

$$\exp[-(\mathbf{a}\cdot\mathbf{n})\, d((x,y), \partial^+\Omega)/\varepsilon], \tag{4.6}$$

where $d((x,y), \partial^+\Omega)$ denotes the distance from the point (x,y) to the outflow boundary. Thus in cross-section perpendicular to $\partial^+\Omega$ these layers are very similar to the boundary layers that we met in one dimension. Their first-order derivatives in the direction perpendicular to the boundary have magnitude $\mathcal{O}(1/\varepsilon)$, and the width of the layer (i.e., the distance one must travel from the boundary before all first-order derivatives are bounded by some constant C) is $\mathcal{O}(\varepsilon \ln(1/\varepsilon))$; recall Remark 3.30.

Layers along $\partial^0\Omega$ are called *parabolic* or *characteristic boundary layers.* In asymptotic expansions of u, these layers can be written as the solution of a parabolic PDE but not as the solution to an ODE; thus they have a much more complicated structure than exponential boundary layers. Their first-order derivatives in the direction perpendicular to the boundary are

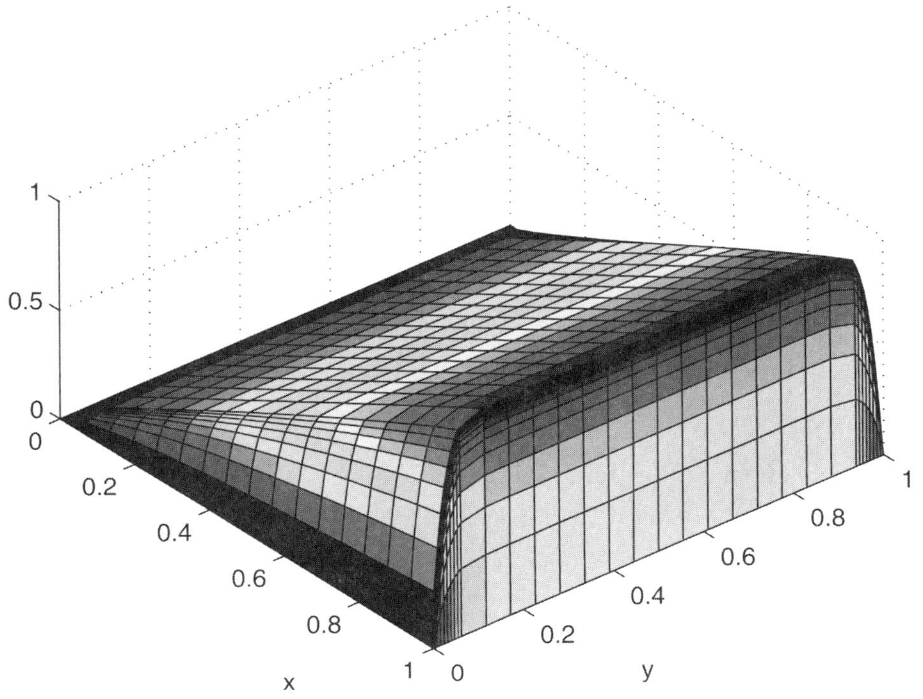

Figure 4.2. Exponential boundary layer and two characteristic boundary layers

$\mathcal{O}(1/\sqrt{\varepsilon})$—not as large as for exponential layers, so characteristic layers are less steep—but the width of the layer is $\mathcal{O}(\sqrt{\varepsilon}\ln(1/\varepsilon))$, so they are wider than exponential layers.

We shall return to this comparison of the two types of layers in Exercise 4.12.

Example 4.2. In Figure 4.2 we plot the solution $u(x,y)$ to the problem

$$-\varepsilon\Delta u(x,y) + u_x(x,y) = 1 \text{ on } \Omega := (0,1)\times(0,1), \quad u(x,y) \equiv 0 \text{ on } \partial\Omega,$$

where $\varepsilon = 0.001$.

The inflow boundary $\partial^-\Omega$ is the side $x = 0$ of $\bar{\Omega}$; the tangential flow boundary comprises the sides $y = 0$ and $y = 1$; the outflow boundary is the remaining side $x = 1$.

From (4.4) each subcharacteristic is parametrized by $x'(t) = 1$, $y'(t) = 0$, so we can take $x = t$, and the subcharacteristics are the lines $y = k$ for arbitrary constant k. Then by (4.5) the reduced problem u_0, written as a function of the parameter t, satisfies $u_0'(t) = 1$, with initial data $u_0(0) = 0$. Hence, $u_0(t) = t$, i.e., $u_0(x,y) = x$ for all $(x,y) \in \Omega$.

On most of Ω one therefore has $u(x,y) \approx x$. The side $x = 1$ of $\bar{\Omega}$ is the outflow boundary $\partial^+\Omega$ and an exponential layer appears there. The tangential flow boundaries $y = 0$ and $y = 1$ have characteristic boundary layers that grow in strength as x moves from 0 to 1 because (e.g., along $y = 0$) $|u_0(x,0) - u(x,0)| = x$ is nonzero and increases as x increases. Observe how the characteristic boundary layer along $y = 0$ is less steep but wider than the exponential layer along $x = 1$.

To determine the asymptotic structure of the solution u inside these layers, one uses stretched variables as in section 2.1. Let us consider first the exponential outflow layer along the side $x = 1$: define the stretched variable $\xi = (1-x)/\varepsilon^p$, where p is a constant to be determined, and rewrite the differential operator in terms of $\tilde{u}(\xi, y) := u(x,y)$, obtaining

$$-\varepsilon^{1-2p}\tilde{u}_{\xi\xi} - \varepsilon\tilde{u}_{yy} - \varepsilon^{-p}\tilde{u}_{\xi} = 0, \tag{4.7}$$

where, as in section 2.1, we replace the right-hand side by zero. To determine the boundary layer function from (4.7), we need an operator that is independent of ε when ε is near zero and will yield a solution $v(\xi, y)$ of (4.7) that decays to zero as $\xi \to \infty$ (i.e., as one moves away from $x = 1$). A consideration of different values of p in (4.7) shows that the only value that yields the correct behaviour of v is $p = 1$, for which (4.7) becomes $-\varepsilon^{-1}v_{\xi\xi} - \varepsilon v_{yy} - \varepsilon^{-1}v_{\xi} = 0$. Letting $\varepsilon \to 0$, this gives $v_{\xi\xi} + v_{\xi} = 0$, with solution $v(\xi, y) = v(\xi, 0)e^{-\xi}$. That is, the layer function at the side $x = 1$ is $v(x,y) = -u_0(1,y)e^{-(1-x)/\varepsilon}$, as $v(\xi, 0) = -u_0(1,y)$ so that $v + u_0 = 0$ (the given boundary value of u) along $x = 1$.

Next, consider the boundary layer along $y = 0$. Define the stretched variable $\eta = y/\varepsilon^q$, where q is a constant to be determined. Changing variables from (x,y) to (x,η), the differential equation for the boundary layer function $w(x,\eta)$ is

$$-\varepsilon w_{xx} - \varepsilon^{1-2q}w_{\eta\eta} - w_x = 0.$$

For ε near zero, only one choice of q yields the desired decay behaviour in w as $\eta \to \infty$: take $q = 1/2$ and w then satisfies $-w_{\eta\eta} - w_x = 0$. This is a parabolic partial differential equation—the heat equation—with initial value $w(x,0)$ chosen to be $-u_0(x,0)$ so that $u_0 + w$ satisfies the given boundary condition $u = 0$ along $y = 0$. Its solution is (see, e.g., [**Eva10**, Section 2.3])

$$w(x,\eta) = -\sqrt{\frac{2}{\pi}} \int_{s=\eta/\sqrt{2x}}^{\infty} e^{-s^2/2} u_0\left(x - \frac{\eta^2}{2s^2}, 0\right) ds. \tag{4.8}$$

The layer along $y = 1$ is of course similar to the layer along $y = 0$. This example is also discussed in [**RST08**, Example III.1.16].

Assuming that w represents all the characteristic layer along $y = 0$, Exercise 4.3 shows that inside the layer the first-order derivative parallel to the boundary is bounded, but the first-order derivative perpendicular to the boundary is $\mathcal{O}(1/\sqrt{\varepsilon})$.

Exercise 4.3. Use formula (4.8) to show that

$$\frac{\partial w(x,y)}{\partial y} = \sqrt{\frac{2}{\pi}} \int_{s=y/\sqrt{2\varepsilon x}}^{\infty} \frac{y}{\varepsilon s^2} e^{-s^2/2}\, ds \quad \text{for } (x,y) \in (0,1)^2.$$

Deduce that

$$\left| \frac{\partial v_0(x,y)}{\partial y} \right| \leq \begin{cases} \dfrac{C\sqrt{x}}{\sqrt{\varepsilon}} & \text{if } y \leq \sqrt{2\varepsilon x}, \\ \dfrac{C\sqrt{x}}{\sqrt{\varepsilon}} e^{-y^2/(8\varepsilon x)} & \text{if } y > \sqrt{2\varepsilon x}. \end{cases}$$

Show likewise that $|\partial v_0(x,y)/\partial x| \leq C$ for all $(x,y) \in (0,1)^2$.

Exercise 4.4. Let $\Omega := (0,1) \times (0,1)$. Consider the problem

$$-\varepsilon \Delta u(x,y) + y^{\beta_1}(1-y)^{\beta_2} u_x(x,y) = y^{\beta_1}(1-y)^{\beta_2} f(x,y) \quad \text{on } \Omega,$$
$$u(x,y) = 0 \quad \text{on } \partial\Omega,$$

where the constants β_1 and β_2 are nonnegative and the function f is smooth and bounded. Show that the solution of the reduced problem does not necessarily agree with the boundary condition $u = 0$ on the sides $x = 1$ and $y = 0, 1$ of $\bar{\Omega}$; consequently, we expect boundary layers along these sides of $\bar{\Omega}$. To investigate the width of the layer along $y = 0$, define the stretched variable $\eta = y/\varepsilon^p$, where p is a constant to be determined (cf. section 2.1), and rewrite the differential equation in terms of (x, η). Show that this leads to the choice $p = 1/(2+\beta_1)$. Show similarly that one should use the stretched variable $\eta' = (1-y)/\varepsilon^{p'}$ with $p' = 1/(2+\beta_2)$ along $y = 1$, and the stretched variable $\xi = (1-x)/\varepsilon$ along $x = 1$. This exercise is based on an example in [**KO10**].

It can happen that there is no characteristic boundary, so only exponential layers appear in the solution, as in the next example.

Example 4.5. In Figure 4.3 we plot the solution $u(x,y)$ to the problem

$$-\varepsilon \Delta u(x,y) + u_x(x,y) + 2u_y(x,y) = 4 \quad \text{on } \Omega := (0,1) \times (0,1),$$
$$u(x,y) \equiv 0 \quad \text{on } \partial\Omega,$$

where $\varepsilon = 0.001$.

Here $\mathbf{a} = (1,2)$ is never tangential to $\partial\Omega$, so there are no characteristic layers. There are exponential layers at the outflow boundaries $x = 1$ and $y = 1$.

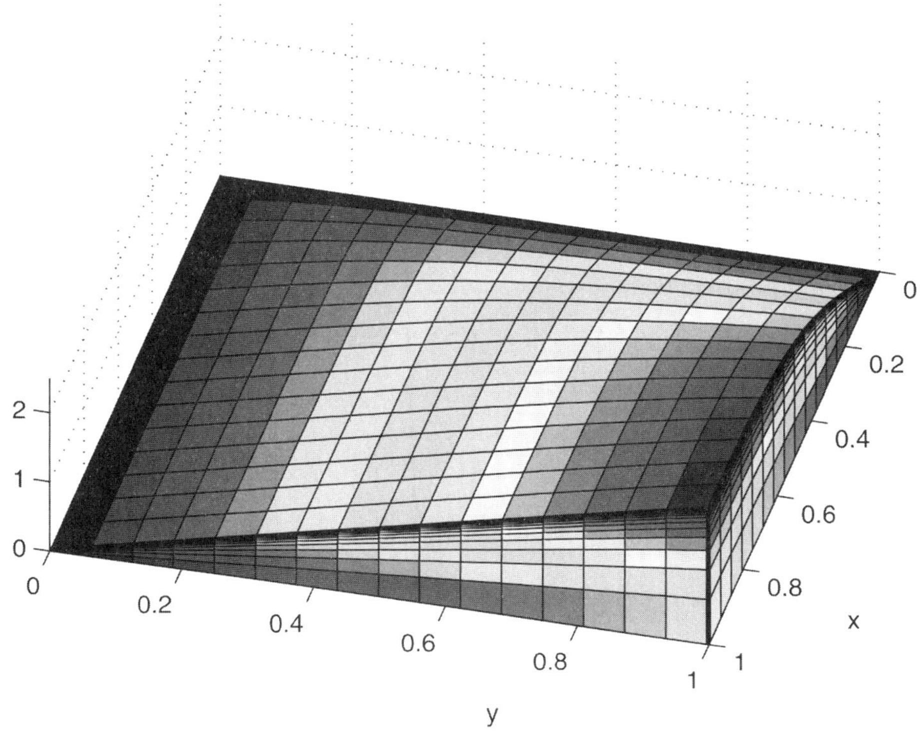

Figure 4.3. Two exponential boundary layers

Problems like Example 4.5, where one does not have to deal with the added complication of characteristic layers, have been a favourite in analyses of numerical methods for convection-diffusion problems.

Exercise 4.6. Compute the reduced solution for Example 4.5 and check that it agrees with Figure 4.3. Find the leading terms for the exponential layers along $x = 1$ and $y = 1$ by imitating the analysis of Example 4.2, then differentiate these terms to get bounds on the derivatives of the solution u inside these layers.

As well as boundary layers, solutions of convection-diffusion problems in two-dimensional domains can have *interior layers* if there is a discontinuity in the boundary data on $\partial^-\Omega$. This phenomenon has no analogue in one-dimensional problems. (The interior layers of Remark 2.42 have a different structure.) From the theory of first-order PDEs, if g has a jump discontinuity at a point $(\hat{x}, \hat{y}) \in \partial^-\Omega$, then u_0 will be discontinuous across the subcharacteristic $\Gamma(\hat{x}, \hat{y})$ that passes through $(\hat{x}, \hat{y})$. Now first-order PDEs preserve Dirichlet boundary data discontinuities but second-order elliptic PDEs smooth out such discontinuities, so the solution $u(x, y)$ of (4.1) will be continuous across $\Gamma(\hat{x}, \hat{y})$. At the same time, u must be close to u_0 once we are a small distance away from $\Gamma(\hat{x}, \hat{y})$. Combining these facts, we deduce that u has an interior layer along the subcharacteristic $\Gamma(\hat{x}, \hat{y})$. Such layers have an asymptotic structure similar to characteristic boundary layers; they are often referred to as *parabolic* or *characteristic interior layers.*

Example 4.7. In Figure 4.4 we use the same differential operator as in Example 4.2, with $\varepsilon = 10^{-6}$. A jump discontinuity has been introduced in the inflow boundary data:

$$g(0, y) = \begin{cases} 0 & \text{for } 0 \le y < 0.5, \\ 1 & \text{for } 0.5 < y \le 1. \end{cases}$$

Consequently, the reduced solution is

$$u_0(x, y) = \begin{cases} x & \text{for } 0 \le y < 0.5, \\ 1 + x & \text{for } 0.5 < y \le 1. \end{cases}$$

This gives rise to an interior layer along the subcharacteristic passing through the discontinuity at $(0.5, 0)$, that is, along the line $y = 0.5$. A homogeneous Dirichlet boundary condition is assumed on the other three sides of the domain, so there are characteristic boundary layers along $y = 0$ and $y = 1$, and an exponential outflow layer at $x = 1$.

Example 4.8. Consider the problem

$$Lu(x, y) := -\varepsilon \Delta u(x, y) + u_x(x, y) + 2u_y(x, y) = 0 \quad \text{on } \Omega := (0, 1) \times (0, 1),$$

where the boundary condition is $u(x, y) = g(x, y)$ with

$$g(x, y) = \begin{cases} 0 & \text{when } y = 0, \\ 1 & \text{otherwise.} \end{cases}$$

There is no tangential flow boundary. The inflow boundary $\partial^-\Omega$ comprises the sides $x = 0$ and $y = 0$ of $\bar{\Omega}$. In (4.5) the functions b and f are both zero, so the reduced solution $u_0(x, y)$ is just the initial data on $\partial^-\Omega$ propagated along the subcharacteristics of L without change. These subcharacteristics are the lines $y = 2x + k$ for arbitrary constant k.

The solution $u(x, y)$ is as usual very close to u_0 away from layers. The outflow boundary $\partial^+\Omega$ comprises the sides $x = 1$ and $y = 1$ of $\bar{\Omega}$. Along the portion $0 \le x \le 1/2$ of the side $y = 1$ there is no layer because $u_0 = g$ there. There are exponential boundary layers along the rest of $\partial^+\Omega$. An interior layer emanates across Ω from the discontinuity in g at the point $(0, 0)$, i.e., along the line $y = 2x$ that is the subcharacteristic through $(0, 0)$; see Figure 4.5, where $\varepsilon = 0.001$.

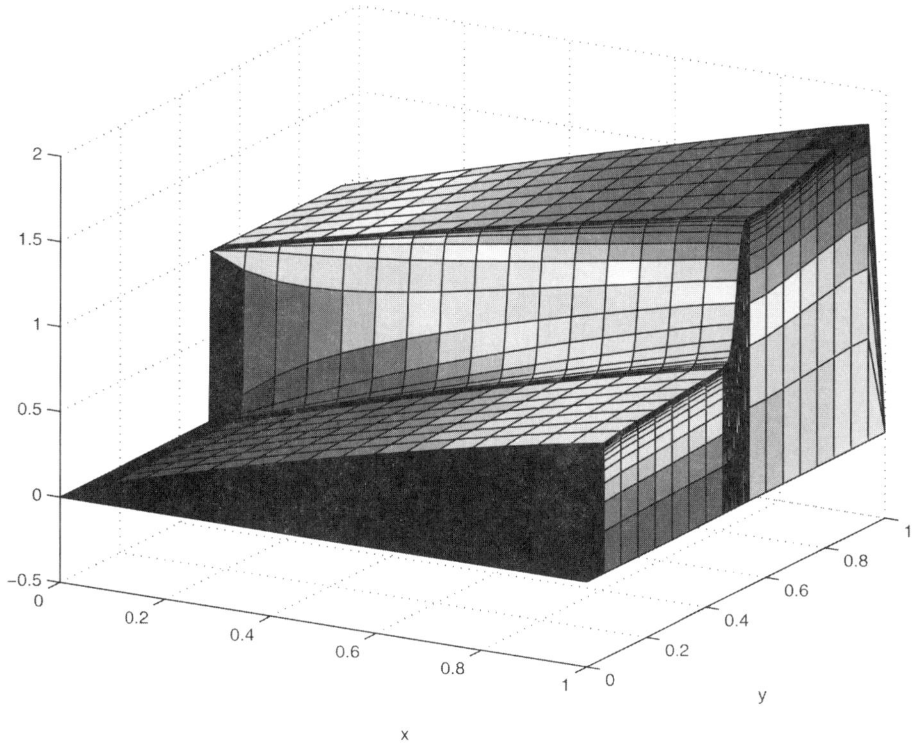

Figure 4.4. Interior layer

The interior layers in the above examples lie along straight lines because in each case the convective coefficient **a** is constant. Any interior layer caused by a discontinuity in the inflow boundary data will follow **a**, so if **a** is variable then the interior layer will be curved; see for example [**MS97**, Figure 13].

Exercise 4.9. Modify the data on the inflow boundary in Exercise 4.1 as follows: $u = 1$ on the side $y = 0$ of Ω, and $u = 2$ on the side $x = 0$. Show that now the reduced solution u_0 has a discontinuity along the curve $y = x^2/2$. Consequently, the solution u of the convection-diffusion problem will exhibit an interior layer along this curve.

Asymptotic expansions of the solutions to several specific cases of (4.1) are given in [**Il′92**].

4.2. A priori estimates

In this section we present various a priori results for the solution of (4.1).

Many analyses in the literature assume the condition

$$\mathbf{a}(x, y) = \big(a_1(x, y), a_2(x, y)\big) > (\alpha_1, \alpha_2) > (0, 0) \quad \text{on } \Omega, \tag{4.9}$$

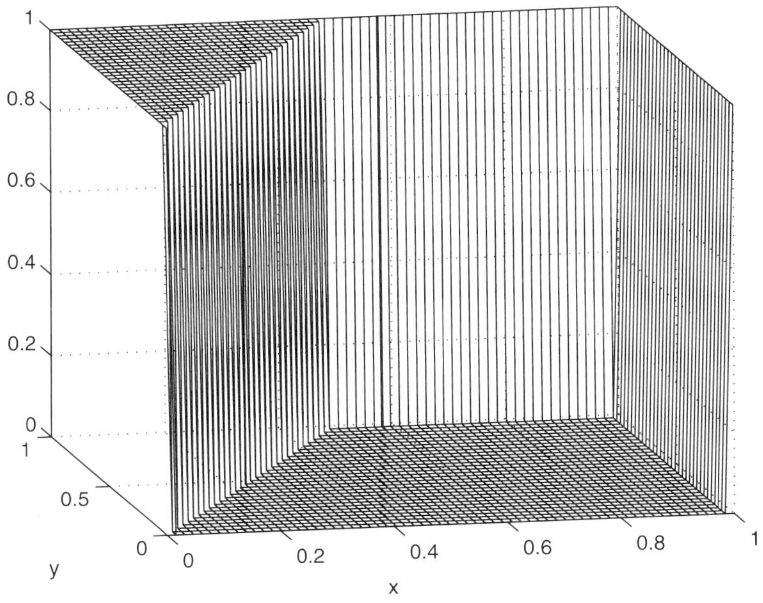

Figure 4.5. Solution of Example 4.8

which in the case where Ω is the unit square $(0,1)^2$ ensures that no characteristic boundary layers are present.

Lemma 4.10. *Assume that* (4.9) *holds true. Then*

(i) *There exists a constant C, which depends on the domain Ω, such that*

$$\|u\|_{L^\infty(\Omega)} \le \|g\|_{L^\infty(\partial\Omega)} + \frac{C\|f\|_{L^\infty(\Omega)}}{\max\{\alpha_1, \alpha_2\}}. \tag{4.10}$$

If Ω is the unit square $(0,1)^2$, then $C = 1$.

(ii) *For each $\delta > 0$, define $\Omega_\delta = \{x \in \Omega : \operatorname{dist}(x, \partial^+\Omega \cup \partial^0\Omega) > \delta\}$. Let $g \in C(\partial\Omega)$. Then there exists a constant $C = C(\delta)$ such that $|u(x,y) - u_0(x,y)| \le C\varepsilon$ for all $(x,y) \in \Omega_\delta$.*

Proof. The proof of (i) is similar to the proof of Lemma 2.14.

The hypotheses of (ii) ensure that there are no interior layers. The proof can be found in [**GFL+83**]. □

Lemma 4.10(ii) states precisely what we have seen in our figures: excluding layers, the solution u is very close to the reduced solution u_0.

Exercise 4.11. Consider (4.1) on $\Omega = (0,1)^2$. Assume that (4.9) holds true and $b \ge 0$ on Ω. Imitate the proof of Lemma 2.14 to show that $|u_x| \le C$

and $|u_y| \le C$ on the inflow boundary $\partial^-\Omega$ of Ω. Thus no layers appear along $\partial^-\Omega$.

Next we discuss the behaviour of derivatives of the solution u of (4.1). Suppose that Ω is the unit square $(0,1)^2$ and the differential operator is as in Example 4.8, so that (4.9) holds true. Then the sides $x=1$ and $y=1$ form the outflow boundary $\partial^+\Omega$. Assuming that no extra complications such as interior layers are present, near $x=1$ one expects the solution u to satisfy the bound

$$\left|\frac{\partial^{i+j}u(x,y)}{\partial x^i \partial y^j}\right| \le C\big(1+\varepsilon^{-i}e^{-(1-x)/\varepsilon}\big), \tag{4.11}$$

while near $y=1$ one expects

$$\left|\frac{\partial^{i+j}u(x,y)}{\partial x^i \partial y^j}\right| \le C\big(1+\varepsilon^{-j}e^{-2(1-y)/\varepsilon}\big). \tag{4.12}$$

Here the multipliers of $1-x$ and $1-y$ correspond to the coefficients of u_x and u_y, respectively, in the definition of Lu; (4.11) and (4.12) are analogues of Exercise 2.22.

Close to the corner (1,1) there will be an *outflow corner layer*, which is like a product of exponential boundary layers, and it satisfies the bound

$$\left|\frac{\partial^{i+j}u(x,y)}{\partial x^i \partial y^j}\right| \le C\big[1+\varepsilon^{-(i+j)}e^{-(1-x)/\varepsilon}e^{-2(1-y)/\varepsilon}\big]. \tag{4.13}$$

Despite the extra negative powers of ε in (4.13), corner layers of this type rarely cause difficulty for numerical methods because they decay so rapidly as one moves away from the corner.

A rigorous proof of bounds such as (4.11)–(4.13) is a delicate and lengthy matter. For problems posed on the unit square $(0,1)^2$, with exponential outflow layers along $x=1$ and $y=1$, it is shown in [**LS01a**] that one has the decomposition $u = S + E_1 + E_2 + E_{12}$, where

$$\left|\frac{\partial^{i+j}S(x,y)}{\partial x^i \partial y^j}\right| \le C, \tag{4.14a}$$

$$\left|\frac{\partial^{i+j}E_1(x,y)}{\partial x^i \partial y^j}\right| \le C\left[1+\varepsilon^{-i}e^{-\alpha_1(1-x)/\varepsilon}\right], \tag{4.14b}$$

$$\left|\frac{\partial^{i+j}E_2(x,y)}{\partial x^i \partial y^j}\right| \le C\left[1+\varepsilon^{-j}e^{-\alpha_2(1-y)/\varepsilon}\right], \tag{4.14c}$$

$$\left|\frac{\partial^{i+j}E_{12}(x,y)}{\partial x^i \partial y^j}\right| \le C\left[1+\varepsilon^{-(i+j)}e^{-\alpha_1(1-x)/\varepsilon}e^{-\alpha_2(1-y)/\varepsilon}\right], \tag{4.14d}$$

for all $(x,y)\in\bar\Omega$ and $0\le i+j\le 3$. Here S is the smooth component of u, E_1 and E_2 are exponential boundary layers, and E_{12} is a corner layer. These bounds are obtained under the extra assumptions that the Dirichlet boundary condition $g(x,y)$ is a continuous function and that a sufficient number of *compatibility conditions* hold true at the corners of $\bar{\Omega}$. See also the discussion in [**KO10**].

Corner compatibility conditions are relationships between the data of the problem and the differential operator that ensure that derivatives of u up to a desired order are continuous on the closed domain $\bar{\Omega}$. They arise only at corners and are not caused by the singularly perturbed nature of the problem. Grisvard [**Gri85**] provides a general exposition of compatibility conditions for elliptic operators on polygonal domains and Han and Kellogg [**HK90**] write down the precise form that they take when applied to convection-diffusion problems posed on the unit square.

If compatibility conditions beyond a certain order are not satisfied at a corner of a domain, then certain derivatives of that order and higher orders must blow up as one approaches this corner. Kellogg and Stynes [**KS05**, **KS07**] consider a problem similar to Example 4.2:

$$-\varepsilon\Delta u + a_1 u_x + bu = f \quad \text{on } \Omega := (0,1)^2, \quad u = g \quad \text{on } \partial\Omega \tag{4.15}$$

where a_1 and b are positive constants. They decompose the solutions of this problem into smooth and layer components; an exponential layer at the outflow boundary $x = 1$, characteristic layers along the tangential flow boundaries $y = 0$ and $y = 1$, and a corner layer at each corner. The bounds in [**KS05**, **KS07**] are expressed in terms of the number of compatibility conditions that are satisfied at each corner of $\bar{\Omega}$. Near $x = 1$, but away from corners, one has (4.11). For the layer component w associated with the characteristic boundary $y = 1$, one finds that

$$\left|\frac{\partial^{i+j} w(x,y)}{\partial x^i \partial y^j}\right| \le C(\sqrt{\varepsilon}\,)^{-j} e^{-k(1-y)/\sqrt{\varepsilon}} \tag{4.16}$$

for a certain positive constant k, provided one stays away from corners. Near the corners, singularities in the derivatives begin to appear; we do not give the details here.

The data of Example 4.8 are not fully compatible at the corner (1,1) with the differential operator L. This incompatibility will cause singularities in the derivatives of u at (1,1). The interaction between these singularities and the exponential and corner layers is not yet fully understood. That is, we are currently unable to write down reliable sharp pointwise bounds on the derivatives of u near the point (1,1), but one expects that sharp bounds are at least as bad as (4.13) and will blow up as (x,y) approaches $(1,1)$.

Comparing the bounds (4.12) and (4.16), we see that derivatives in the direction perpendicular to the boundary are larger inside exponential layers than inside parabolic layers. This is apparent in Figure 4.2 for instance. But this figure also shows that on the other hand, characteristic layers are wider than exponential layers. We now outline how to quantify this "wideness".

Engineers working in fluid dynamics define *boundary layer width* or *boundary layer thickness* to be the distance from the boundary at which a boundary layer component has decreased to 1% of its magnitude at the boundary. A numerical analyst might define the layer width/thickness to be the distance from the boundary beyond which all the first m derivatives of the boundary layer component (for some fixed positive integer m) are bounded independently of ε.

Exercise 4.12. Using the above definitions, show that the boundary layer width/thickness of

(i) an exponential layer component $e^{-k(1-y)/\varepsilon}$ is $\mathcal{O}(\varepsilon|\ln\varepsilon|)$ for a numerical analyst and $\mathcal{O}(\varepsilon)$ for an engineer;

(ii) a characteristic layer component is $e^{-k(1-y)/\sqrt{\varepsilon}}$ is $\mathcal{O}(\sqrt{\varepsilon}\,|\ln\varepsilon|)$ for a numerical analyst and $\mathcal{O}(\sqrt{\varepsilon})$ for an engineer.

Here k is a positive constant. Quantities depending on 1% or m are hidden inside the $\mathcal{O}(\cdot)$ notation but all dependence on ε is explicit.

In this book we focus on boundary layers of width $\mathcal{O}(\varepsilon)$ and $\mathcal{O}(\sqrt{\varepsilon})$ (ignoring $\ln\varepsilon$ factors), but other widths are possible as Exercise 4.4 shows.

It is in general difficult to derive sharp bounds on derivatives of solutions of convection-diffusion problems inside characteristic boundary and interior layers. Although such bounds are of great interest to numerical analysts, few rigorous results appear in the literature. As we described above, pointwise bounds for characteristic boundary layers posed on the unit square are proved in [**KS05**, **KS07**]. In another paper [**KS06**] the same authors consider a convection-diffusion problem in a half-plane with a discontinuity in an arbitrary specified derivative of the boundary data and derive pointwise bounds on derivatives of the solution, including the behaviour along the interior layer emanating from the point of discontinuity.

Exercise 4.13. For a problem posed on the unit square $(0,1)^2$ with a characteristic layer along $y = 0$, one can sometimes prove (see, for example, Exercise 4.3) that the layer component v of the solution satisfies

$$\left|\frac{\partial v(x,y)}{\partial y}\right| \le \frac{C}{\sqrt{\varepsilon}}\, e^{-ky^2/\varepsilon} \quad \text{on } (0,1)^2 \quad \text{for some positive constant } k.$$

Show that this bound implies the slightly weaker but more tractable bound

$$\left|\frac{\partial v(x,y)}{\partial y}\right| \le \frac{C}{\sqrt{\varepsilon}}\, e^{-ky/\sqrt{\varepsilon}} \quad \text{on } (0,1)^2.$$

4.2.1. Sobolev norms. Let $\|\cdot\|_k$ and $|\cdot|_k$ denote the usual norm and seminorm on the Sobolev space $H^k(\Omega)$ for all nonnegative integers k. In particular $\|\cdot\|_0 = \|\cdot\|_{L^2(\Omega)}$.

The presence of layers in u means that one does not have $\|u\|_k \le C$ for any $k \ge 1$. Even in one dimension, the H^k norm of the function $e^{-\alpha(1-x)/\varepsilon}$ is easily checked (see Exercise 2.31) to be $\mathcal{O}\left(\varepsilon^{-k+1/2}\right)$, and exponential layers in two-dimensional problems have a similar magnitude. This observation motivates the following definition of a weighted energy norm that is commonly used in finite element analyses of convection-diffusion problems: for all $w \in H^1(\Omega)$, set

$$\|w\|_{1,\varepsilon} = \sqrt{\varepsilon|w|_1^2 + \|w\|_0^2}.$$

See Section 6.1 for further discussion of this norm.

Lemma 4.14. *Let u be the solution of (4.1). Assume that $b - (\operatorname{div}\mathbf{a})/2 \ge C_5 > 0$ on $\bar{\Omega}$ for some constant C_5. Assume also that Ω is convex or has smooth boundary. Then there exists a constant C such that*

$$\varepsilon^{3/2}|u|_2 + \varepsilon^{1/2}|u|_1 + \|u\|_0 \le \varepsilon^{3/2}|u|_2 + \sqrt{2}\,\|u\|_{1,\varepsilon} \le C.$$

Proof. Let G be the solution of the problem $\Delta G = 0$ on Ω, $G = g$ on $\partial\Omega$. Then the hypotheses on the domain Ω ensure that $\|G\|_2 \le C\|g\|_{0,\partial\Omega}$ by a classical inequality (see, e.g., [**GT01**]). Subtract G from u to reduce the problem to the case of homogeneous Dirichlet boundary conditions: setting $\tilde{u} := u - G$, we have $\tilde{u} = 0$ on $\partial\Omega$ and $L\tilde{u} = \tilde{f}$ on Ω, where $\tilde{f} := f + \varepsilon\Delta G - \mathbf{a}\cdot\nabla G - bG$.

Now use a standard energy norm argument: multiply $L\tilde{u} = \tilde{f}$ by $\tilde{u}$, then integrate by parts, obtaining

$$\varepsilon|\tilde{u}|_1^2 + \int_\Omega \left(b - \frac{1}{2}\operatorname{div}\mathbf{a}\right)\tilde{u}^2 = \int_\Omega \tilde{f}\tilde{u} \le \|\tilde{f}\|_0\|\tilde{u}\|_0 \le \frac{1}{2C_5}\|\tilde{f}\|_0^2 + \frac{C_5}{2}\|\tilde{u}\|_0^2,$$

and $\|\tilde{u}\|_{1,\varepsilon} \le C$ follows, where $C = C(\|\tilde{f}\|_0) = C(\|f\|_0, \|g\|_{0,\partial\Omega})$ because $\|G\|_2 \le C\|g\|_{0,\partial\Omega}$. Then by a triangle inequality we get

$$\|u\|_{1,\varepsilon} \le \|\tilde{u}\|_{1,\varepsilon} + \|G\|_{1,\varepsilon} \le C, \tag{4.17}$$

where C depends on $\|f\|_0$ and $\|g\|_{0,\partial\Omega}$.

The PDE (4.1) and (4.17) now yield

$$\varepsilon\|\Delta u\|_0 \le C(|u|_1 + \|u\|_0 + \|f\|_0) \le C(\varepsilon^{-1/2} + 1) \le C\varepsilon^{-1/2},$$

so $\varepsilon^{3/2}\|\Delta u\|_0 \le C$. But the classical inequality $|u|_2 \le C(\|\Delta u\|_0 + \|u\|_0)$ holds true [**GT01**], and we get $\varepsilon^{3/2}|u|_2 \le C$. □

Remark 4.15. Analogously to Remark 2.12, if (4.9) holds true then one can assume without loss of generality that $b - (\operatorname{div}\mathbf{a})/2 \ge C_5 > 0$ on $\bar{\Omega}$.

Exercise 4.16. For all $(x, y) \in (0,1)^2$ and $0 \le i + j \le 1$, suppose that the function w satisfies the bound (4.16). Thus w represents a typical characteristic boundary layer along $y = 0$. Compute $\|w\|_{1,\varepsilon}$ and show that if $\varepsilon \to 0$, then $\|w\|_{1,\varepsilon} \to 0$ also. This says that the standard energy norm is unsuited to measuring the strength of characteristic boundary layers because its ε-weighting is too strong—unlike the situation for exponential boundary layers that is explored in Exercise 4.17.

Exercise 4.17. For $\Omega = (0,1)^2$, compute $\|w\|_{1,\varepsilon}$ for

(i) $w(x, y) = w_1(x, y) := e^{-\alpha(1-x)/\varepsilon}$,

(ii) $w(x, y) = w_{12}(x, y) := e^{-\alpha(1-x)/\varepsilon} e^{-\beta(1-y)/\varepsilon}$,

where α, β are positive constants. Show that $\|w_1\|_{1,\varepsilon} = \mathcal{O}(1)$ as $\varepsilon \to 0$; this means that the norm $\|\cdot\|_{1,\varepsilon}$ is weighted correctly for exponential boundary layers such as w_1. Is it weighted correctly for the corner layer w_{12}?

4.2.2. Some other observations. Dörfler [**Dör99**] gives bounds on u and its derivatives in various norms (both isotropic and anisotropic) and for a variety of convection-diffusion problems on bounded domains. Pointwise bounds on derivatives of u for many variants of (4.1) are derived in [**SS09**] but the arguments are sometimes presented in a very concise style.

The derivation of asymptotic expansions and bounds on derivatives of solutions in two-dimensional domains can be difficult. In [**Hem96**] Hemker considers the following model problem on the exterior of the unit disc:

$$\begin{aligned} -\varepsilon\Delta u + u_x &= 0 \text{ on } \mathbb{R}^2 \setminus D, \\ u(x, y) &= 1 \quad \text{on } \partial D, \\ u(x, y) &\to 0 \quad \text{as } (x, y) \to \infty, \end{aligned}$$

where $D := \{(x, y) \in \mathbb{R}^2 : x^2 + y^2 \le 1\}$. Here the flow and boundary conditions generate a boundary layer that has a very complicated structure near the points $(0, \pm 1)$ on ∂D. He derives an exact formula for the solution u by writing it as an infinite series of modified Bessel functions, but this is impractical for evaluating the solution accurately, so he goes on to examine asymptotic expansions of the solution in various parts of $\mathbb{R}^2 \setminus D$. Subsequently, various numerical methods have been used in attempts to solve this difficult problem accurately; see the webpage `http://homepages.cwi.nl/~pieth/webs/webs.html`

Remark 4.18. Consider the reaction-diffusion problem

$$-\varepsilon\Delta u + bu = f \quad \text{on } \Omega := (0,1)^2, \quad u = g \text{ on } \partial\Omega,$$

where $b(x, y) \ge 2\beta^2 > 0$ on $\bar{\Omega}$. Analogously to one-dimensional problems of this type (see Remarks 2.37 and 2.50), the solution u exhibits typically a boundary layer on all four sides of $\bar{\Omega}$. Assuming some compatibility conditions at the corners of the domain, Clavero et al. [**CGO05**] decompose the solution as $u = S + \sum_{i=1}^{4} w_i + \sum_{i=1}^{4} z_i$, where S is smooth, each w_i is a boundary layer associated with one side of $\bar{\Omega}$, and each z_i is a corner layer associated with one corner of $\bar{\Omega}$. For example, it is shown that the layer component w_1 associated with the side $y = 1$ satisfies

$$(4.18) \qquad \left| \frac{\partial^j w_1(x,y)}{\partial y^j} \right| \le C(\sqrt{\varepsilon}\,)^{-j} e^{-\beta(1-y)/\sqrt{\varepsilon}} \quad \text{for } j = 0, 1, \ldots, 4.$$

This is a natural generalization of one-dimensional reaction-diffusion problems; each of these boundary layers is exponential. Andreev [**And06**] investigates what happens to these bounds when the corner compatibility conditions are not satisfied.

Observe that the bound (4.18) is similar to the bound (4.16) for characteristic boundary layers. This happens because in the problem (4.15), near the boundary $y = 1$, the term $a_1 u_x$ is not large so the PDE behaves like $-\varepsilon\Delta u + bu = f - a_1 u_x$ with a bounded right-hand side, i.e., a reaction-diffusion problem! Nevertheless one should not think that characteristic boundary layers are the same as reaction-diffusion boundary layers, for characteristic layers have a more complicated structure and the bound (4.16) is a simplification of the true state of affairs; recall Exercise 4.13.

Exercise 4.19. Consider the one-dimensional reaction-diffusion problem $-\varepsilon v'' + bv = r$ on $(0,1)$, with $v(0)$ and $v(1)$ given. Assuming sufficient smoothness of b and r, the decomposition $v = S + E_0 + E_1$ of the solution into its smooth and layer components is given in Remark 2.50. Show that $\|E_j\|_{1,\varepsilon} = \mathcal{O}(\varepsilon^{1/4})$ for $j = 0, 1$, and consequently $\|v - S\|_{1,\varepsilon} \to 0$ as $\varepsilon \to 0$. This says that, with respect to the norm $\|\cdot\|_{1,\varepsilon}$, the reaction-diffusion problem is regularly perturbed in the sense of the commuting diagram (1.5); it is *not* singularly perturbed. Use the same decomposition of v to prove that the reaction-diffusion problem is singularly perturbed with respect to the $L^\infty[0,1]$ norm.

Exercise 4.20. By using the decomposition of Theorem 2.44 for the solution of the one-dimensional convection-diffusion problem (2.14), show that this problem is singularly perturbed with respect to the norm $\|\cdot\|_{1,\varepsilon}$.

For further discussion of the material in this section, see [**MOS12**], [**RST08**, Section III.1], and the references therein.

4.3. General comments on numerical methods

Numerical methods (such as central differencing on equidistant meshes) that contain no mechanism for stabilizing solutions in exponential layers will usually have wild oscillations in their computed solutions on much of Ω, as in section 3. As we shall see, this problem can be handled by modifying the discretisation of the convective terms (e.g., using some form of finite difference upwinding or special choices of finite element trial and test spaces) and by modifying the mesh (e.g., a two-dimensional Shishkin mesh). When this is done correctly, one can compute accurate solutions inside these layers.

Characteristic layers, on the other hand, differ in both respects:

- If the method has no stabilizing mechanism specifically designed to address characteristic layers and no special mesh is used for these layers, then the layer will induce small oscillations in the computed solution. But these oscillations usually appear only inside and near the characteristic layer, so the solution can still be computed accurately on the rest of Ω.
- It is often difficult—at least in the case of interior layers—to compute accurate solutions inside characteristic layers.

Thus one could use some form of upwinding (i.e., some discrete approximation of $\mathbf{a}\cdot\nabla u$ that is skewed away from the outflow boundary) to stabilize the method for exponential layers, combined with some heuristic mesh refinement near characteristic layers. Whether or not the mesh refinement yields an accurate solution inside the characteristic layers, nevertheless the solution elsewhere will be accurate.

The following pair of examples is related to our observation that one can to a certain extent neglect characteristic layers but not exponential layers, and it is also related to Exercises 4.16 and 4.17.

Consider again Example 4.8 but with $g(x,y) \equiv 1$. Then the solution $u(x,y)$ has exponential boundary layers along $x = 1$ and $y = 1$. The reduced solution $u_0(x,y)$ will of course ignore these layers, and one finds that $\|u - u_0\|_{1,\varepsilon} = \mathcal{O}(1)$.

On the other hand the solution u of Example 4.2 has two characteristic boundary layers and one exponential boundary layer. Schieweck [**Sch86**] proves that if one sets $v(x,y) = u_0(x,y) - u_0(1,y)e^{-(1-x)/\varepsilon}$ (this is the reduced solution plus an appropriate exponential layer term, so v ignores only the parabolic layers), then $\|u - v\|_{1,\varepsilon} \le C\varepsilon^{1/4}$.

Chapter 5

Finite Difference Methods in Two Dimensions

Consider the boundary value problem (4.1) posed on the unit square Ω and under the hypothesis (4.9) on the convection coefficients, viz.,

$$\mathbf{a}(x,y) = \big(a_1(x,y), a_2(x,y)\big) > (\alpha_1, \alpha_2) > (0,0) \quad \text{on } \Omega.$$

Assume that the mesh $\{(x_i, y_j)\}$ is rectangular and equidistant in each coordinate direction: $x_i = ih$ and $y_j = jk$ for $i = 0, \ldots, N$ and $j = 0, \ldots, M$ with $h := 1/N$ and $k := 1/M$.

We use a standard approximation of the second-order derivatives:

(5.1)

$$u_{xx}(x_i, y_j) \approx \frac{u^N_{i+1,j} - 2u^N_{ij} + u^N_{i-1,j}}{h^2}, \quad u_{yy}(x_i, y_j) \approx \frac{u^N_{i,j+1} - 2u^N_{ij} + u^N_{i,j-1}}{k^2},$$

where u^N_{ij} is the computed solution at each mesh point (x_i, y_j).

5.1. Extending one-dimensional approaches

As for one-dimensional problems, approximating the first-order derivatives in (4.1) by central differences

$$u_x(x_i, y_j) \approx \frac{u^N_{i+1,j} - u^N_{i-1,j}}{2h} \quad \text{and} \quad u_y(x_i, y_j) \approx \frac{u^N_{i,j+1} - u^N_{i,j-1}}{2k}$$

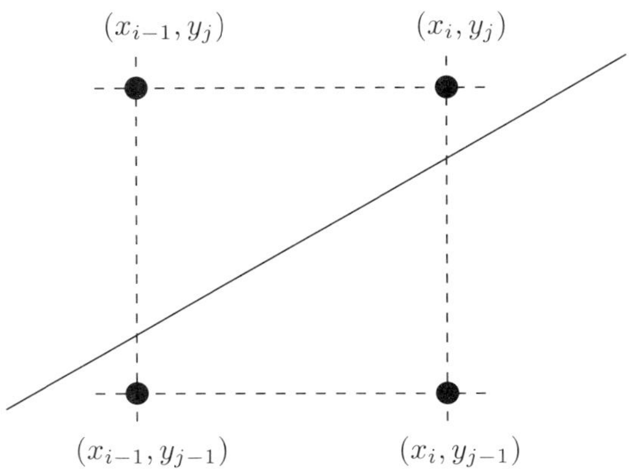

Figure 5.1. Mesh points and line indicating nearby interior layer

leads to an unstable method. Instead, one can use simple upwinding,

$$u_x(x_i, y_j) \approx \frac{u_{i,j}^N - u_{i-1,j}^N}{h} \quad \text{and} \quad u_y(x_i, y_j) \approx \frac{u_{i,j}^N - u_{i,j-1}^N}{k},$$

and this yields an M-matrix (we leave this as an exercise!). Combining this with (5.1) and the approximation $u(x_i, y_j) \approx u_{ij}^N$ for the zero-order term in (4.1), the resulting method is stable, but we expect from our experience with ODEs that it will smear exponential boundary layers.

In fact, one can foresee heuristically that this method will also smear interior layers. In Figure 5.1, the solid line indicates an interior layer in the solution. Now the value of $u(x_i, y_j)$ depends strongly on the u values along the upstream portion of the subcharacteristic that passes through (x_i, y_j)—this is a line through (x_i, y_j) parallel to the solid line drawn—but simple upwinding makes $u(x_i, y_j)$ depend on $u(x_i, y_{j-1})$, which introduces inaccuracies because the value of $u(x_i, y_j)$ has little to do with the values of u on the other side of the interior layer.

A difference scheme on a family of arbitrary rectangular meshes is said to be *robust* or *uniformly convergent (with respect to* ε*) of order* $\beta > 0$ *in the discrete* L^∞ *norm* if its solution $\{u_{ij}^N\}$ satisfies $|u_{ij} - u_{ij}^N| \le CN^{-\beta}$ for $i, j = 0, \dots, N$ and all sufficiently small H, independently of ε. Here we take $N+1$ mesh points in each coordinate direction for simplicity, H is the mesh diameter, β is some positive constant that is independent of the mesh and of ε, and we write u_{ij} instead of $u(x_i, y_j)$ (we shall do likewise for all other functions in $C(\bar{\Omega})$).

For uniform convergence on an equidistant mesh, an analogue of Theorem 3.18 shows that once again the coefficients in the scheme must have a certain exponential character [**RST08**, p.265]. One can define a five-point scheme that is a two-dimensional analogue of the Il′in–Allen–Southwell scheme of Example 3.22. When the data of (4.1) are smooth, the convective term satisfies the separability condition $\mathbf{a}(x, y) = (a_1(x), a_2(y))$, one has (4.9), and some compatibility conditions are satisfied at the corners of Ω, this scheme can be proved to achieve uniform convergence of order 1 in the discrete L^∞ norm [**RS15a**], like its one-dimensional analogue in Example 3.22. Nevertheless this scheme, which is a form of upwinding, can smear interior layers quite badly.

See also [**Gos13**], where the finite difference generalisations of Il′in, Allen, and Southwell to balance laws are considered (recall our Remark 3.9).

Remark 5.1. In the one-dimensional case, when the convective coefficient and right-hand side of the differential equation are constants and the reaction term is zero, one can construct a three-point scheme (the Il′in–Allen–Southwell scheme) that computes the true solution exactly at each mesh point; see Exercise 3.23. This construction is possible because the difference scheme must be satisfied exactly by only three functions: a particular solution of the differential equation and two linearly independent solutions of the homogenous differential equation. But in two dimensions no analogous result is possible for a scheme that uses a fixed number of points because the homogeneous differential equation has infinitely many linearly independent solutions.

5.2. Shishkin meshes

Continuing in the footsteps of our earlier sections, we now consider a two-dimensional Shishkin mesh for the problem (4.1) on the unit square, while assuming that (4.9) is satisfied so that the solution has exponential boundary layers along $x = 1$ and $y = 1$. Let N, an even integer, be the number of mesh intervals in each coordinate direction. Define the transition points on the x- and y-axes to be $1 - \lambda_x$ and $1 - \lambda_y$, respectively, where $\lambda_x = (2\varepsilon/\alpha_1)\ln N$ and $\lambda_y = (2\varepsilon/\alpha_2)\ln N$. The fine and coarse mesh regions on the coordinate axes each contain $N/2$ mesh intervals. See Figure 5.2 for the mesh with $N = 8$, which shows the tensor product of the one-dimensional Shishkin mesh of section 3.4 with itself.

One can define simple upwinding on this rectangular two-dimensional Shishkin mesh by applying the one-dimensional formula (3.20) in each coordinate direction, as follows. Set $h_i = x_i - x_{i-1}$ for each i and $k_j = y_j - y_{j-1}$ for each j. For each mesh function $\{v_{i,j}\}_{i,j=0}^N$, set

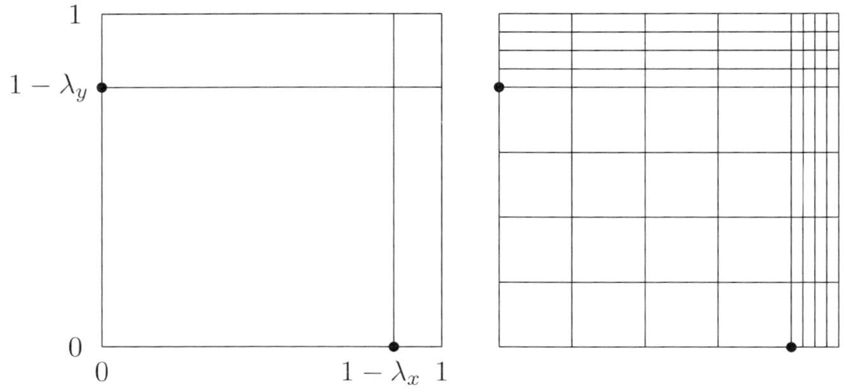

Figure 5.2. Shishkin mesh with $N = 8$ for two exponential outflow layers

$D_x^- v_{ij} = (v_{i,j} - v_{i-1,j})/h_i$, $D_y^- v_{ij} = (v_{i,j} - v_{i,j-1})/k_j$, and

$$\delta_x^2 v_{ij} = \frac{2}{h_i + h_{i+1}} \left(\frac{v_{i+1,j} - v_{i,j}}{h_{i+1}} - \frac{v_{i,j} - v_{i-1,j}}{h_i} \right),$$

$$\delta_y^2 v_{ij} = \frac{2}{k_j + k_{j+1}} \left(\frac{v_{i,j+1} - v_{i,j}}{k_{j+1}} - \frac{v_{i,j} - v_{i,j-1}}{k_j} \right).$$

The two-dimensional simple upwind difference scheme for approximating (4.1) is then

$$-\varepsilon(\delta_x^2 + \delta_y^2) u_{ij}^N + a_1(x_i, y_j) D_x^- u_{ij}^N + a_2(x_i, y_j) D_y^- u_{ij}^N + b_{ij} u_{ij}^N = f_{ij} \quad \text{for } i, j = 1, \ldots, N-1,$$

$$u_{i,j}^N = g_{ij} \text{ if } \{i, j\} \cap \{0, N\} \text{ is nonempty.}$$

This is a five-point scheme. Its associated matrix is an M-matrix.

Assuming that the decomposition and bounds (4.14) for the true solution are valid, an analysis similar to that of section 3.4 shows that the simple upwind solution u_{ij}^N on a Shishkin mesh satisfies

$$(5.2) \qquad |u_{ij} - u_{ij}^N| \le C N^{-1} \ln N \quad \text{for all } i, j.$$

That is, one gets almost first-order uniform convergence in the discrete L^∞ norm.

Exercise 5.2. Consider the problem (4.1), with $\Omega = (0, 1)^2$, under the hypothesis (4.9). Assume the decomposition (4.14). This problem is solved numerically using simple upwinding on a rectangular Shishkin mesh $\{(x_i, y_j)\}$ of the same type as Figure 5.2. Show that the matrix associated with this difference scheme is an M-matrix. Use this property and the given decomposition of u to prove (5.2). *Hint.* Imitate closely the proof of Theorem 3.39. You will find that the one-dimensional barrier functions defined there can be used in two dimensions also.

If one modifies this scheme (as in the hybrid difference scheme of Exercise 3.42) by using central differencing instead of simple upwinding wherever the Shishkin mesh is fine in the relevant coordinate direction, then the M-matrix property is retained and a variant of the simple upwind analysis yields (see [**LS99**]) the improved bound

$$|u_{ij} - u_{ij}^{N,\text{hybrid}}| \le CN^{-1} \quad \text{for all } i, j,$$

where $u_{ij}^{N,\text{hybrid}}$ is the solution computed by this hybrid scheme.

Kopteva [**Kop03**] shows, under some extra compatibility assumptions at the corners, that one iteration of Richardson extrapolation applied to the simple upwind solution u_{ij}^N on the Shishkin mesh yields a solution v_{ij}^N for which

$$|u_{ij} - v_{ij}^N| \le CN^{-2}(\ln N)^2 \quad \text{for all } i, j.$$

Approximation of the first-order derivatives of u is also discussed in Koptera's paper.

5.3. Characteristic boundary layers

We now discard the hypothesis that $\mathbf{a}(x, y) > (0, 0)$ on Ω in order to introduce characteristic boundary layers into the problem.

Remark 5.3 (Shishkin's obstacle theorem)**.** The convergence results earlier in Chapter 5 are all proved under hypotheses that exclude characteristic layers. The difficulty of accurately approximating characteristic boundary layers is underlined by a remarkable negative result of Shishkin [**Shi89**], which we now describe. Suppose the solution of the problem has a characteristic boundary layer. Suppose also that one applies any difference scheme on an equidistant mesh whose coefficients are drawn from a fixed class of functions (e.g., the Il′in–Allen–Southwell scheme, whose coefficients are all exponentials and polynomials; the point is that one is forbidden to vary the difference scheme by choosing the type of coefficients to correspond exactly to the precise nature of each new set of boundary data). Then *this scheme cannot yield uniform convergence of any positive order in the discrete L^∞ norm inside the characteristic boundary layer for all smooth and compatible boundary data g.* The essential reason for this negative result is that, at each point (x, y) near $\partial^0\Omega$, a characteristic boundary layer depends on all the data along that connected component of $\partial^0\Omega$ (see (4.8)). This is quite unlike an exponential boundary layer, whose behaviour at (x, y) near $\partial^+\Omega$ depends only on the difference between the reduced solution u_0 and the

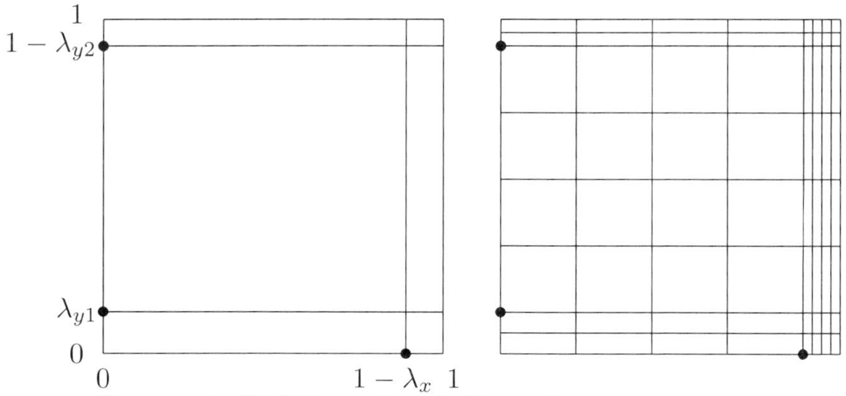

Figure 5.3. Shishkin mesh with $N = 8$ for one exponential and two characteristic layers (see Example 4.2)

boundary data at the nearest boundary point (see (4.6)), a much simpler situation.[1]

For a detailed proof of this result in the context of a time-dependent problem, cf. [**MOS12**, Chapter 15].

Consider now the problem (4.1) with $\mathbf{a} = (a_1, 0)$ where $a_1 > \alpha_1 > 0$, and $\Omega = (0,1)^2$. The solution of this problem (cf. Example 4.2) has an exponential boundary layer at $x = 1$ and characteristic boundary layers at $y = 0$ and $y = 1$. A Shishkin mesh appropriate to this problem is constructed as follows. Use an x-axis transition point exactly as in Figure 5.2. Place y-axis transition points at λ_{y1} and $1 - \lambda_{y2}$, where each λ_{yk} is $\mathcal{O}\left(\varepsilon^{1/2} \ln N\right)$, then use $N/4$ equidistant mesh intervals in each of $[0, \lambda_{y1}]$ and $[1 - \lambda_{y2}, 1]$ and $N/2$ equidistant mesh intervals in $[\lambda_{y1}, 1 - \lambda_{y2}]$; see Figure 5.3. For simple upwinding, O'Riordan and Shishkin [**OS08**] show that under certain fairly strong hypotheses on the smoothness and compatibility of the data of the problem, simple upwinding on this Shishkin mesh yields

$$|u_{ij} - u_{ij}^N| \le CN^{-1}(\ln N)^2 \quad \text{for all } i, j,$$

where u_{ij}^N is the computed solution. This difference scheme is stabilised in the direction of flow by the use of simple upwinding (this addresses the exponential outflow layer) but the scheme contains no mechanism to stabilise the characteristic boundary layers, which are handled entirely by the Shishkin mesh.

A large collection of numerical computations on Shishkin meshes for various problems can be found in [**FHM+00**]. The construction and implementation of Shishkin meshes for boundary layers along straight portions

[1] *Historical Note.* The "obstacle" described in Remark 5.3 motivated Grisha Shishkin to search for an approach other than fitted schemes to obtain uniform convergence inside characteristic layers. His solution? The Shishkin mesh.

of $\partial\Omega$ is straightforward once the asymptotic nature of the layer has been ascertained, and most numerical examples in the literature are of this type. Shishkin meshes for curved boundary layers were used in [**Kop07b**]; see also [**KO10**]. For curved interior layers there are few examples in the literature; see [**MS97**] for a heuristic approximation of a Shishkin meshes that yields a visually satisfactory solution.

Shishkin and Shishkina's book [**SS09**] contains a wealth of theoretical results for finite difference methods applied on these meshes to solve many convection-diffusion problems. A partial summary of these is given in [**KO10**].

5.4. Other remarks

Remark 5.4 (Defect correction method)**.** This is a general technique that seeks to generate a useful higher-order finite difference scheme for any problem by combining a stable low-order scheme with a higher-order but unstable scheme.

Consider an arbitrary rectangular mesh. Compute an initial approximation $\hat{u}^N$ by using simple upwinding: solve $L_{\mathrm{up}}^N \hat{u}^N = f^N$. Substitute this solution into the formally higher-order central difference scheme L_c^N to compute the "defect" $\sigma^N := f^N - L_c^N \hat{u}^N$. Then compute the "defect correction" δ^N by solving $L_{\mathrm{up}}^N \delta^N = \sigma^N$. Hence, we form the final solution $u^N := \hat{u}^N + \delta^N$.

This method avoids instability by solving only discrete systems that involve the upwind operator L_{up}^N, yet aims to attain the higher-order convergence associated with the operator L_c^N. The idea can be placed in a more general setting and has been applied to many problems unrelated to convection-diffusion; see [**BR84**]. For analyses of a defect correction method that combines simple upwinding and central differencing for one-dimensional convection-diffusion problems, see [**Lin04**] and [**LK10**]. Defect correction is related to Richardson extrapolation, and to obtain a rigorous proof of its validity in two dimensions on a Shishkin mesh like that of Figure 5.2 would require some extension of the delicate two-dimensional analysis in [**Kop03**]. Nevertheless numerical results for the method are encouraging; see Remark 6.50.

Finally, we point out that in convection-diffusion problems, when one no longer assumes hypotheses such as $\mathbf{a}(\cdot,\cdot) > (0,0)$, then although simple upwinding remains stable (i.e., the computed solution is free of nonphysical oscillations), it can give dangerously misleading results. Brandt and Yavneh [**BY91**] give an example of linearized recirculating flow in an annulus where the subcharacteristics are circles and, except near the boundary of the domain, the solution computed by a version of simple upwinding is $\mathcal{O}(1)$ distant from the true solution!

Remark 5.5. For the two-dimensional reaction-diffusion problem of Remark 4.18, one can use a Shishkin mesh that is a tensor product of the one-dimensional Shishkin meshes of Remark 3.44 together with a standard five-point discretisation of the differential equation. For N mesh intervals in each coordinate direction, this scheme is shown by Clavero et al. [**CGO05**] to yield a computed solution $\{u_{ij}^N\}$ that satisfies

$$|u_{ij} - u_{ij}^N| \le C(N^{-1} \ln N)^2 \quad \text{for all } i, j, \tag{5.3}$$

under suitable smoothness and corner compatibility conditions on the data of the problem. Andreev [**And06**] uses a sophisticated argument to discard these compatibility conditions, though his bound $|u_{ij} - u_{ij}^N| \le CN^{-2}(\ln N)^4$ is slightly worse than (5.3).

Chapter 6

Finite Element Methods

If one attempts to solve a convection-diffusion problem by means of a standard Galerkin finite element method (FEM) with linear or bilinear elements on an equidistant mesh, then a typical computed solution will display large oscillations. Thus some mechanism is needed to stabilize an FEM: a special choice of trial or test functions, or a special mesh, or a modification of the standard bilinear form, or a combination of these devices. In the sections that follow we discuss each in turn.

Two good books on finite element methods are [**BS08**] and [**GT17**].

6.1. The loss of stability in the (Bubnov–)Galerkin FEM

In the standard Galerkin finite element method—which more precisely could be called the Bubnov–Galerkin FEM—the trial and test spaces are the same, except perhaps for the boundary conditions on each; see [**BS08**].

To begin, we return to one-dimensional problems. Recall the convection-diffusion two-point boundary value problem (3.1):

$$Lu(x) := -\varepsilon u''(x) + a(x)u'(x) + b(x)u(x) = f(x) \quad \text{for } 0 < x < 1, \tag{6.1a}$$

$$u(0) = u(1) = 0, \tag{6.1b}$$

where $0 < \varepsilon \leq 1$, $a(x) \geq \underline{a} > \alpha > 0$, and $b(x) \geq 0$ on [0,1], and a, b, and f lie in $C^1[0,1]$.

To solve (6.1) numerically on an equidistant mesh $x_i = i/N$ (for $i = 0, 1, \ldots, N$), suppose we use a standard Galerkin method with piecewise polynomials. Let $S^N = \text{span}\{\phi_i\}_{i=1}^{N-1}$ denote the trial space. Then the computed solution $u^N(x) := \sum_{i=1}^{N-1} u^N(x_i)\phi_i(x) \in S^N$ is generated by a weak form of the differential equation:

$$\int_0^1 \left[\varepsilon(u^N)'(x)\phi_i'(x) + a(x)(u^N)'(x)\phi_i(x) + b(x)u^N(x)\phi_i(x)\right] dx \\ = \int_0^1 f(x)\phi_i(x)\,dx \quad \text{for } i = 1, \ldots, N-1. \tag{6.2}$$

Exercise 6.1. For the two-point boundary value problem (6.1) with $a(\cdot)$ constant and $b \equiv 0$, show that the Galerkin FEM with piecewise linears on an equidistant mesh will generate the same discrete matrix as the central differencing method (3.2).

Exercise 6.1 implies that when one applies the piecewise linear Galerkin FEM to solve Example 1.1, one obtains again the unsatisfactory oscillatory solution of Figure 3.1. In Chapter 3 we explained this phenomenon by means of M-matrices and discrete L^∞-norm arguments, but in a finite element context, it is more natural to work with $L^2[0,1]$ and related Sobolev norms, so we shall do this now.

The standard norm used in finite element analyses of second-order two-point boundary value problems on $[0,1]$ is the Sobolev $H^1[0,1]$ norm

$$\|v\|_1 = \left(|v|_1^2 + \|v\|_0^2\right)^{1/2}, \tag{6.3}$$

where $\|\cdot\|_0$ is the $L^2[0,1]$ norm and $|v|_1 = \|v'\|_0$ is the $H^1[0,1]$ seminorm. Typical solutions of (6.1) contain a layer component $z(x) := e^{-k(1-x)/\varepsilon}$ for some constant $k > 0$, and a quick calculation shows that $|z|_1 = \mathcal{O}(\varepsilon^{-1/2})$. This implies that the norm (6.3) is scaled incorrectly for measuring solutions of (6.1); it should be replaced by

$$\|v\|_{1,\varepsilon} := \left(\varepsilon|v|_1^2 + \|v\|_0^2\right)^{1/2}, \tag{6.4}$$

so that both $\varepsilon^{1/2}|u|_1$ and $\|u\|_0$ are $\mathcal{O}(1)$.

Assume that

$$b(x) - \frac{a'(x)}{2} \geq C_5 > 0 \quad \text{for } x \in [0,1] \text{ and some constant } C_5. \tag{6.5}$$

(Since $a(\cdot) > \alpha > 0$, one can always obtain (6.5) by a change of variable, as in Remark 2.12.) Multiply the differential equation (6.1a) by u then integrate over $[0,1]$; after an integration by parts, one has

$$\varepsilon|u|_1^2 + \int_0^1 \left(b - \frac{a'}{2}\right) u^2\,dx = \int_0^1 f u\,dx.$$

Hence

$$\varepsilon|u|_1^2 + C_5\|u\|_0^2 \leq \|f\|_0\|u\|_0 \leq \frac{1}{2C_5}\|f\|_0^2 + \frac{C_5}{2}\|u\|_0^2,$$

using the Cauchy–Schwarz and arithmetic-geometric mean inequalities. Thus

$$\varepsilon|u|_1^2 + \frac{C_5}{2}\|u\|_0^2 \le \frac{1}{2C_5}\|f\|_0^2,$$

which implies the *stability bound* (cf. Lemma 4.14)

$$\|u\|_{1,\varepsilon} \le C\|f\|_0 \quad \text{for some constant } C. \tag{6.6}$$

Inequality (6.6) is a sharp a priori estimate of the true solution u of (6.1), as $\varepsilon^{1/2}|u|_1$, $\|u\|_0$, and $\|f\|_0$ are all typically $\mathcal{O}(1)$. But what analogous sharp a priori bound should we expect for an accurate piecewise linear approximation u_h of u that is computed by a finite element method? The answer "$\|u_h\|_{1,\varepsilon} \le C\|f\|_0$" is incorrect: for if u_h were a piecewise linear interpolant of u on an equidistant mesh of diameter $h \gg \varepsilon$, then

$$\int_0^{1-h} \varepsilon(u_h')^2\,dx = \mathcal{O}(\varepsilon)$$

and, because of the boundary layer in u at $x = 1$,

$$\int_{1-h}^1 \varepsilon(u_h')^2\,dx = \varepsilon\int_{1-h}^1 \mathcal{O}\left(\frac{1}{h^2}\right)\,dx = \mathcal{O}\left(\frac{\varepsilon}{h}\right),$$

so

$$\varepsilon^{1/2}|u_h|_1 = \mathcal{O}\left(\varepsilon/h\right)^{1/2} \quad \text{but } \|u_h\|_0 = \mathcal{O}(1) \text{ and } \|f\|_0 = \mathcal{O}(1),$$

i.e., the scaling of $\varepsilon^{1/2}|u_h|_1$ makes this term excessively small.

Nevertheless this calculation also shows that $h^{1/2}|u_h|_1$ gives the correct scaling (i.e., yields a term that is $\mathcal{O}(1)$) when u_h is the piecewise linear interpolant of u. Thus we seek finite element methods whose computed solutions v_h satisfy the stability bound

$$h^{1/2}|v_h|_1 + \|v_h\|_0 \le C\|f\|_0, \tag{6.7}$$

with a constant C that is independent of h and ε.

Exercise 6.2. Here is an alternative motivation for the weighted seminorm $h^{1/2}|v_h|_1$. Let u be the solution of (6.1). Let v be any function that satisfies $v(1) = u(1) = 0$ and $v(1-h) = u(1-h)$, where $1 > h \gg \varepsilon$, so typically one has $v(1-h) = \mathcal{O}(1)$. Now

$$|v(1-h)| = \left|\int_{1-h}^1 v'(x)\,dx\right|.$$

Apply the Cauchy–Schwarz inequality to bound this integral. In general, when does this inequality give a sharp bound? Use this property to justify the weighting $h^{1/2}|v|_1$ for v that is linear on $[1-h, 1]$. Why doesn't this justify the weighting $h^{1/2}|v|_1$ for $v = u$ on $[1-h, 1]$?

The oscillating piecewise linear solution v_h of the standard Galerkin FEM that is displayed in Figure 3.1 has

$$h\int_{0.75}^{1}(v_h')^2\,dx \approx Ch\int_{0.75}^{1}\left(\frac{1}{h}\right)^2 dx = Ch^{-1} \not\leq C\|f\|_0^2,$$

so it does not satisfy the bound (6.7). Thus (6.7) excludes large nonlocal oscillations like those of Figure 3.1; this is why we refer to it as a stability bound.

Exercise 6.3. Suppose that the standard Galerkin FEM with piecewise polynomials is used to solve (6.1) on an equidistant mesh. Deduce from (6.2) that its solution u^N satisfies

$$\varepsilon^{1/2}|u^N|_1 + \|u^N\|_0 \leq C\|f\|_0 \quad \text{for some constant } C.$$

This stability bound is of course weaker than (6.7), and as Figure 3.1 shows, it is not strong enough to forbid damaging oscillations in the computed solution.

Remark 6.4. In general, stabilized FEMs that are satisfactory for convection-diffusion problems satisfy stability bounds that may look different from (6.7) but turn out to have some connection with it. This connection is obvious for the streamline diffusion FEM of section 6.4; in the case of continuous interior penalty stabilization, see Remark 6.55.

Remark 6.5. If instead of piecewise linears, one uses higher-degree piecewise polynomials in the Galerkin FEM, this has a certain stabilizing effect; see section 6.5 and [**BR94**, **CKSL+14**, **KT11**].

6.2. Relationship to classical FEM analysis

Consider once more the two-dimensional boundary problem (4.1):

(6.8a)
$$Lu(x,y) := -\varepsilon\Delta u(x,y) + \mathbf{a}(x,y)\cdot\nabla u(x,y) + b(x,y)u(x,y) = f(x,y)$$

on a bounded domain $\Omega \subset \mathbb{R}^2$, with

$$u(x,y) = 0 \quad \text{on } \partial\Omega, \tag{6.8b}$$

where $0 < \varepsilon \leq 1$, and the functions $\mathbf{a}, b$, and f are assumed to be Hölder continuous on $\bar{\Omega}$, the closure of Ω. For convenience, in (6.8b) we took the Dirichlet boundary condition to be homogeneous. We shall assume (cf. Remark 4.15) that

$$b(x,y) - \frac{\operatorname{div}\mathbf{a}(x,y)}{2} \geq C_5 > 0 \quad \text{on } \bar{\Omega} \text{ for some constant } C_5, \tag{6.9}$$

as is often done in classical finite element analyses.

Write $(\cdot,\cdot)$ for the $L_2(\Omega)$ innner product, i.e., $(f,g) := \int_\Omega fg$. Then the $L_2(\Omega)$ norm of a function f is $\|f\|_0 := (f,f)^{1/2}$, its $H^1(\Omega)$ seminorm is $|f|_1 := (\nabla f, \nabla f)$, and its $H^1(\Omega)$ norm is $\|f\|_1 = (\|f\|_0^2 + |f|_1^2)^{1/2}$. The Sobolev space $H^1(\Omega)$ is the space of functions v defined on Ω for which $\|v\|_1$ is finite. Its subspace $H_0^1(\Omega)$ comprises those functions in $H^1(\Omega)$ whose traces vanish on the boundary $\partial\Omega$.

The standard weak form of the boundary value problem (6.8) is to find $u \in H_0^1(\Omega)$ such that

$$B(u,w) = (f,w) \quad \text{for all } w \in H_0^1(\Omega), \tag{6.10}$$

where the bilinear form $B(\cdot,\cdot)$ is defined by

$$B(v,w) := (\varepsilon\nabla v, \nabla w) + (\mathbf{a}\cdot\nabla v, w) + (bv, w) \quad \forall v,w \in H^1(\Omega). \tag{6.11}$$

Now (6.9) implies the *coercivity/stability inequality*

$$B(v,v) \geq \min\{1, C_5\}\|v\|_{1,\varepsilon}^2 \quad \forall v \in H_0^1(\Omega), \tag{6.12}$$

where

$$\|v\|_{1,\varepsilon} := (\|v\|_0^2 + \varepsilon|v|_1^2)^{1/2}.$$

This inequality is essentially equivalent to (6.6), as can be seen by comparing the derivations of (6.6) and (6.12).

Suppose now that (6.10) is discretised using an FEM with globally continuous piecewise linears on a triangulation of Ω. One can then prove, as for (6.12), that the (Bubnov–)Galerkin FEM solution u_{Gal}^N satisfies the coercivity inequality $B(u_{\text{Gal}}^N, u_{\text{Gal}}^N) \geq \min\{1, C_5\}\|u_{\text{Gal}}^N\|_{1,\varepsilon}^2$, but as we saw in section 6.1 this property is not strong enough to exclude bad oscillations from u_{Gal}^N. Instead, we need some two-dimensional analogue of the stronger property (6.7) for our FEM solution to be accurate.

There is a further difficulty that arises when trying to extend classical FEM analyses to convection-diffusion problems: as well as stability, one needs the property of continuity or boundedness of the bilinear form [**BS08**, Section 2.5]. Now the bilinear form $B(\cdot,\cdot)$ of (6.11) satisfies the coercivity inequality (6.12) with respect to the norm $\|\cdot\|_{1,\varepsilon}$, so the classical argument requires boundedness of $B(\cdot,\cdot)$ with respect to the same norm, i.e., it requires an inequality of the form

$$|B(v,w)| \leq C\|v\|_{1,\varepsilon}\|w\|_{1,\varepsilon} \quad \text{for all } v,w \in H_0^1(\Omega), \tag{6.13}$$

where C is some constant (independent of v, w and of course ε). It is easy to see that the diffusion and reaction terms in (6.11) pose no difficulties for (6.13). But a little experimentation with pen and paper will convince the reader that it is impossible to prove (6.13) because of the convection term in (6.11).

Exercise 6.6. Demonstrate that one cannot prove (6.13) with a constant C that is independent of ε, v, and w.

Remark 6.7. For finite element methods, reaction-diffusion problems

$$-\varepsilon\Delta u + bu = f \quad \text{on } \Omega, \quad u = 0 \text{ on } \partial\Omega, \tag{6.14}$$

where $b(x, y) \geq 2\beta^2 > 0$ on $\bar{\Omega}$, are a very different animal from convection-diffusion problems, because their associated bilinear form satisfies the stability inequality (6.12) *and* the continuity bound (6.13). Consequently, the solution u^N_{Gal} computed by any standard Galerkin method with globally continuous trial and test space $V^N \subset H^1_0(\Omega)$ will satisfy the quasi-optimal bound

$$\|u - u^N_{\text{Gal}}\|_{1,\varepsilon} \leq C \inf_{v^N \in V^N} \|u - v^N\|_{1,\varepsilon} \tag{6.15}$$

for some constant C, by Ceá's Lemma [**BS08**, Theorem 2.8.1].

This apparent success is due to the weakness of the norm $\|\cdot\|_{1,\varepsilon}$ when solving reaction-diffusion problems—recall Exercise 4.19! When ε is very small, $\|u\|_{1,\varepsilon} \approx \|u\|_0$ for typical solutions of (6.14), so (6.15) says merely that we get convergence in the $L^2(\Omega)$ norm. As the $L^2(\Omega)$ norm is so weak, no special method is needed to get convergence, unlike, for example, the $L^\infty(\Omega)$ norm, for which (even in one dimension) Remark 3.27 tells us that some special scheme is needed. To address this deficiency of $\|\cdot\|_{1,\varepsilon}$, in [**LS12**] it is replaced by the *balanced norm* $(\|v\|_0^2 + \varepsilon^{1/2}|v|_1^2)^{1/2}$ where the exponent of ε has been changed so that both components of the norm are $\mathcal{O}(1)$ for typical solutions of (6.14) when ε is small.

See Remark 6.38 for further comments on the reaction-diffusion problem.

Exercise 6.8. Verify that the bilinear form associated with (6.14) satisfies the stability inequality (6.12) and the continuity bound (6.13). Deduce that the FEM solution satisfies (6.15).

6.3. L^*-splines

In Chapter 6 our main interest is in FEMs whose solutions satisfy (6.7) or some related inequality, but in the present section we make a detour to consider a class of FEMs that are connected with the famous Il′in–Allen–Southwell scheme of Example 3.22.

To generate this scheme when solving the one-dimensional problem (6.1), we shall use a Petrov–Galerkin FEM, that is, the trial space S^N and test space T^N are not identical, unlike standard (Bubnov–)Galerkin methods.

On the equidistant mesh $x_i = i/N$, for $i = 0, 1, \ldots, N$, the trial space S^N is the standard space of piecewise linear "hat" functions that vanish at

$x = 0, 1$, so the boundary conditions in (6.1b) are satisfied. Let $\bar{a}, \bar{b}$, and $\bar{f}$ be some piecewise-constant approximations of a, b, and f on our mesh. Define the test space T^N to be the space of approximate L^* *splines* spanned by $\{\psi_i\}_{i=1}^{N-1}$, i.e.,

$$\bar{L}^*(\psi_i)(x) := -\varepsilon\psi_i''(x) - \bar{a}(x)\psi_i'(x) + \bar{b}(x)\psi_i(x) = 0 \tag{6.16}$$

on each subinterval (x_{j-1}, x_j), with $\psi_i(x_j) = \delta_{ij}$, the discrete Kronecker delta. Then each ψ_i has support $[x_{i-1}, x_{i+1}]$; see Figure 6.1 for an example.

The computed solution $u^N(x) = \sum_{i=1}^{N-1} u^N(x_i)\phi_i(x) \in S^N$ is generated, as usual in FEMs, by a weak form of the differential equation

$$\begin{aligned}\int_0^1 &\left[\varepsilon(u^N)'(x)\psi_i'(x) + \bar{a}(x)(u^N)'(x)\psi_i(x) + \bar{b}(x)u^N(x)\psi_i(x)\right] dx \\ &= \int_0^1 \bar{f}(x)\psi_i(x)\, dx \qquad \text{for } i = 1, \ldots, N-1.\end{aligned}$$

If one defines $\bar{a}$ by the quadrature rule

$$\int_0^1 \bar{a}(x)(u^N)'(x)\psi_i(x)\, dx = a_i \int_0^1 (u^N)'(x)\psi_i(x)\, dx,$$

with analogous definitions for $\bar{b}$ and $\bar{f}$, then one obtains the Il′in–Allen–Southwell scheme.

Exercise 6.9. Verify that the FEM just described does generate the Il′in–Allen–Southwell scheme.

The alternative choice

$$\bar{a}\Big|_{(x_{j-i}, x_j)} = \frac{a_{j-1} + a_j}{2} \qquad \text{for each } j$$

(with similar definitions for $\bar{b}$ and $\bar{f}$) yields the El Mistikawy–Werle scheme of section 3.3.

Both of these are successful schemes, and the only special construction we made when generating them in an FEM context was to use L^* splines. Why do L^* splines make such good test functions?

The explanation is to be found by considering Green's functions for the differential operator L, which we examined in section 2.2.

For each mesh point $x_i \in (0, 1)$, let $G(\cdot, x_i)$ denote the Green's function associated with that point. (An explicit formula for G in the case of

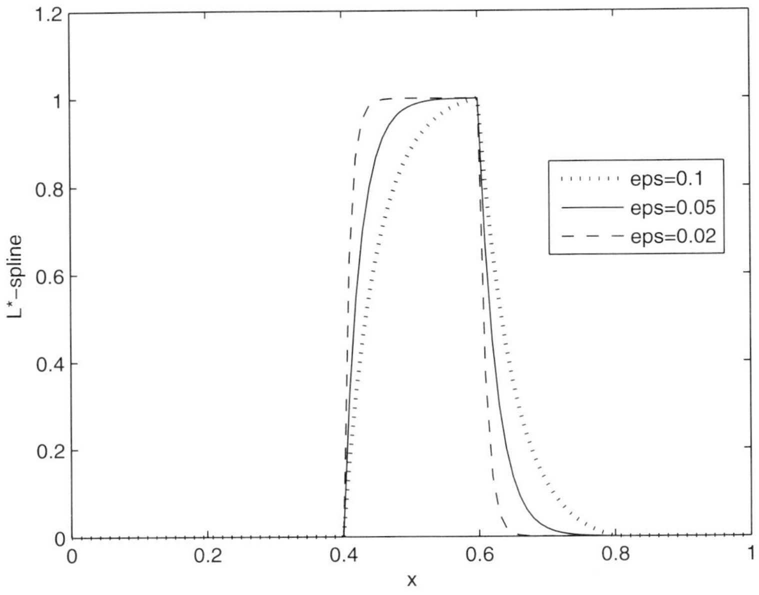

Figure 6.1. L^*-spline ψ_i where $N = 5, x_i = 0.6$ and $\bar{a} = 2, \bar{b} = 0$

constant a and $b \equiv 0$ was given in equation (2.12).) Then

$$\begin{aligned} u_i &= \int_0^1 f(\xi)G(\xi, x_i)\, d\xi \\ &= \int_0^1 (Lu)(\xi)G(\xi, x_i)\, d\xi \\ &= \int_0^1 \left[\varepsilon(u'(\xi))G_\xi(\xi, x_i) + a(x)u'(\xi)G(\xi, x_i) + bu(\xi)G(\xi, x_i)\right] d\xi. \end{aligned}$$

Note the resemblance between this identity and the weak form of the differential equation that was used above to generate the FEM, and note also the similarity between the definitions of G and ψ_i. The key idea of this Petrov–Galerkin FEM is to choose the ψ_i in such a way that the test space T^N is capable of producing a good approximation of the Green's function.

Remark 6.10. When piecewise linears or bilinears are used as the trial space for convection-diffusion problems in one or two dimensions, useful numerical methods on general meshes are based on some test space that is constructed to approximate the Green's function of the continuous operator. This Green's function is skewed away from the outflow boundary; see [**FK12**, **Mor96**] for discussions of its properties in two dimensions.

Remark 6.11. An alternative approach to using test functions that approximate L^* splines is to shift the special construction from the test space to the trial space by using trial functions ϕ that are approximate L splines (i.e., to satisfy some approximate version of $L\phi = 0$), together with some standard space of test functions, such as piecewise linears. The relationship between this dual approach and the use of L^*-spline test functions is discussed at length in [**RST08**, Section I.2.2.5].

Some authors have generalized the L^* splines of (6.16) to two dimensions by taking their tensor product on rectangular grids (see, e.g., [**OS91**]), but this method is applicable only on domains whose boundary comprises straight-line segments, each of which is parallel to one of the coordinate axes, and so negates one of the main advantages of finite element methods over finite differences. Consequently, we do not discuss this approach here but refer the reader to [**RST08**, Section III.3.5.1].

An alternative generalization that is genuinely two dimensional is found in Sacco et al. [**SGG99**]: To solve (4.1) with $b \equiv 0$ on an arbitrary triangular mesh, one uses a trial space with local basis

$$1, \quad e^{(\bar{a}_1 x + \bar{a}_2 y)/\varepsilon}, \quad \bar{a}_1 y - \bar{a}_2 x,$$

where $(\bar{a}_1, \bar{a}_2)$ is a piecewise-constant approximation of $\mathbf{a} = (a_1, a_2)$. Here the functions 1 and $e^{\bar{a}_1 x + \bar{a}_2 y}$ are two-dimensional analogues of the functions that appear in approximate L splines for the corresponding one-dimensional problem (6.1), but the third function $\bar{a}_1 y - \bar{a}_2 x$ is intrinsically two dimensional. Observe that all three functions lie in the null space of the operator $-\varepsilon\Delta(\cdot) + \bar{a}_1(\cdot)_x + \bar{a}_2(\cdot)_y$, i.e., they are approximate L splines. Piecewise linears are used in the test space. It is shown in [**SS98**] that this method is essentially equivalent to the unusual exponentially upwinded scheme used in the software package PLTMG.

A different approach is used in [**BPP15**], which gives an FEM extension of the Il′in–Allen–Southwell formula to unstructured grids in two and three dimensions.

Remark 6.12. In [**DG11**] and subsequent papers, the authors develop a general theory of "optimal test functions" that in the one-dimensional case includes L^* splines as a special case. This paper also provides references to some attempts of other authors to extend the one-dimensional L^*-spline approach to two dimensions. The methodology of [**DG11**] uses the framework of discontinuous Galerkin FEMs.

6.4. The streamline-diffusion finite element method (SUPG)

We now return to FEMs for solving (6.8) that enjoy stability bounds that are two-dimensional analogues of (6.7). Of these, the best known is the *streamline-diffusion FEM* (SDFEM) of Hughes and Brooks [**HB79**]; it is also called the *streamline upwind Petrov-Galerkin method* (SUPG). Although this is one of the earliest FEMs designed specifically for convection-diffusion problems (it dates from 1979), it is still one of the best; see the numerical comparison of various FEMs in [**ACF+11**].

Given a partition Ω^N of Ω into elements τ (triangles or rectangles), let S^N be a conforming trial space (i.e., $S^N \subset H_0^1(\Omega)$) of piecewise polynomials of degree $k \geq 1$ defined on Ω^N. Then the *SDFEM solution* $u_{SD} \in S^N$ of (6.8) is defined by

$$B_{SD}(u_{SD}, w^N) = (f, w^N) + \sum_{\tau \in \Omega^N} \delta_\tau (f, \mathbf{a} \cdot \nabla w^N)_\tau \quad \forall w^N \in S^N, \tag{6.17}$$

where

$$B_{SD}(\cdot,\cdot) := B(\cdot,\cdot) + B_{\text{stab}}(\cdot,\cdot),$$
$B(\cdot,\cdot)$ is the standard Galerkin bilinear form defined in (6.11),

and

$$B_{\text{stab}}(v, w) := \sum_{\tau \in \Omega^N} \delta_\tau (-\varepsilon \Delta v + \mathbf{a} \cdot \nabla v + bv, \, \mathbf{a} \cdot \nabla w)_\tau.$$

Here $(\cdot,\cdot)_\tau$ is the $L_2(\tau)$ inner product and the user-chosen parameter δ_τ is a nonnegative constant on each element $\tau \in \Omega^N$; it will be used to stabilize the method. If $\delta_\tau = 0$ for all $\tau \in \Omega^N$, then the SDFEM reverts to the standard Galerkin method.

The term $\sum_{\tau \in \Omega^N} \delta_\tau (f, \mathbf{a} \cdot \nabla w^N)$ is included in the right-hand side of (6.17) to give the standard FEM property of *Galerkin orthogonality:*

$$B_{SD}(u - u_{SD}, w^N) = 0 \quad \forall w^N \in S^N. \tag{6.18}$$

This identity is also known as the *Galerkin projection property.* It says that the SDFEM is consistent, in an FEM sense, with the boundary value problem (6.8).

Remark 6.13 (Terminology). In the particular case where S^N comprises piecewise linears or bilinears (i.e., $k = 1$ and $S^N = P_1$ or Q_1 in the usual notation for FEM trial spaces [**BS08**]), and $b \equiv 0$, the bilinear form of (6.17) simplifies to

$$\begin{aligned} &B_{SD}(u_{SD}, w^N) \\ &\quad = (\varepsilon \nabla u_{SD}, \nabla w^N) + (\mathbf{a} \cdot \nabla u_{SD}, w^N) + \sum_{\tau \in \Omega^N} \delta_\tau (\mathbf{a} \cdot \nabla u_{SD}, \, \mathbf{a} \cdot \nabla w^N)_\tau, \end{aligned}$$

which is the same as the standard Galerkin bilinear form $B(\cdot,\cdot)$ associated with the differential operator $-\varepsilon \Delta u - \delta |\mathbf{a}|^2 u_{\mathbf{aa}} + \mathbf{a} \cdot \nabla u$, where δ is a piecewise-constant function and $u_{\mathbf{a}}$ denotes the directional derivative in the subcharacteristic direction. That is, we have added artificial diffusion to the PDE, but only in the direction of the subcharacteristics, which for stationary problems are the same as the so-called *streamlines* of the differential operator. This is the explanation of the name SDFEM.

Alternatively, the SDFEM can be regarded as a Petrov–Galerkin method with trial space S^N and test space $\{w^N + \sum_{\tau\in\Omega^N} \delta_\tau \mathbf{a}\cdot\nabla w^N : w^N \in S^N\}$, i.e., the test functions are obtained by "upwinding" the trial functions along the subcharacteristics/streamlines. For this reason it is also known as the SUPG method.

Remark 6.14 (Mesh terminology)**.** Consider a mesh Ω^N that partitions a bounded domain $\Omega \subset \mathbb{R}^2$. Let τ be any triangle (or convex quadrilateral) that lies in Ω^N. Denote the diameter of τ by h_τ, and by ρ_τ the diameter of the largest circle that can be inscribed in τ. Clearly, $h_\tau \geq \rho_\tau$ for all τ. Set $h_N = \max_{\tau\in\Omega_N} h_\tau$.

A family of meshes $\{\Omega^N\}_{N=1,2,\dots}$ is called *shape-regular* if for each N there is a positive constant C, independent of N, such that $h_\tau/\rho_\tau \leq C$ for all $\tau \in \Omega^N$. This condition excludes long thin elements from each mesh in the family; thus, for example, Shishkin meshes are not shape-regular.

A shape-regular family $\{\Omega^N\}$ of meshes is said to be *quasi-uniform* if there is a positive constant C', independent of N, such that for each Ω^N one has $h_\tau \geq C'h_N$ for all $\tau \in \Omega^N$. This condition says that for each Ω^N the diameters of the mesh elements do not vary much over the domain Ω.

Assume that our mesh is quasi-uniform. Then (see, e.g., [**BS08**]) on each element $\tau \in \Omega^N$ one can define an interpolating polynomial u^I of degree at most k that has the standard interpolation property

$$|u - u^I|_{m,\tau} \leq C h_\tau^{k+1-m} |u|_{k+1,\tau} \quad \text{for } m = 0, 1, 2, \tag{6.19}$$

and all members of S^N satisfy the inverse inequality

$$\|\Delta w^N\|_{0,\tau} \leq C_{\text{inv}} h_\tau^{-1} |w^N|_{1,\tau} \quad \forall w^N \in S^N, \tag{6.20}$$

where the $|\cdot|_{\ell,\tau}$ are local Sobolev seminorms on the element τ, the norm on $L_2(\tau)$ is $\|\cdot\|_{0,\tau}$, and h_τ denotes the diameter of τ.

We now define a norm that is stronger than $\|\cdot\|_{1,\varepsilon}$ and natural for the analysis of the SDFEM. For each $v \in H^1(\Omega)$, set

$$\|v\|_{SD} = \Big(\varepsilon|v|_1^2 + \sum_{\tau\in\Omega^N} \delta_\tau \|\mathbf{a}\cdot\nabla v\|_{0,\tau}^2 + C_5\|v\|_0^2\Big)^{1/2}. \tag{6.21}$$

Lemma 6.15. *Suppose that the mesh is quasi-uniform and the SDFEM parameter δ_τ satisfies*

$$0 \le \delta_\tau \le \frac{1}{2} \min\left\{ \frac{C_5}{\|b\|^2_{L_\infty(\tau)}}, \frac{h_\tau^2}{\varepsilon C_{\text{inv}}^2} \right\} \quad \textit{for each } \tau \in \Omega^N. \tag{6.22}$$

Then the bilinear form $B_{SD}(\cdot,\cdot)$ is coercive with respect to $\|\cdot\|_{SD}$ over $S^N \times S^N$, i.e.,

$$B_{SD}(w^N, w^N) \ge \frac{1}{2}\|w^N\|^2_{SD} \quad \forall w^N \in \Omega^N.$$

Proof. For each $w^N \in \Omega^N$, one gets easily

$$\begin{aligned} B_{SD}(w^N, w^N) &\ge \varepsilon|w^N|_1^2 + C_5\|w^N\|_0^2 + \sum_{\tau\in\Omega^N} \delta_\tau\|\mathbf{a}\cdot\nabla w^N\|^2_{0,\tau} \\ &\quad + \sum_{\tau\in\Omega^N} \delta_\tau(-\varepsilon\Delta w^N + bw^N, \mathbf{a}\cdot\nabla w^N)_\tau. \end{aligned} \tag{6.23}$$

Now the inequality $st \le s^2 + t^2/4$ for s and $t \ge 0$, the inverse inequality (6.20) and the hypothesis (6.22) on δ_τ yield

$$\begin{aligned} &\Big|\sum_{\tau\in\Omega^N} \delta_\tau(-\varepsilon\Delta w^N + bw^N, \mathbf{a}\cdot\nabla w^N)_\tau\Big| \\ &\quad\le \sum_{\tau\in\Omega^N} \Big[\varepsilon^2\delta_\tau\|\Delta w^N\|^2_{0,\tau} + \delta_\tau\|b\|^2_{L_\infty(\tau)}\|w^N\|^2_{0,\tau} + \frac{1}{2}\delta_\tau\|\mathbf{a}\cdot\nabla w^N\|^2_{0,\tau}\Big] \\ &\quad\le \frac{1}{2}\Big[\varepsilon|w^N|_1^2 + C_5\|w^N\|_0^2 + \sum_{\tau\in\Omega^N} \delta_\tau\|\mathbf{a}\cdot\nabla w^N\|^2_{0,\tau}\Big]. \end{aligned}$$

Applying this bound in (6.23) then gathering similar terms, the lemma is proved. □

Thus the SDFEM is coercive on the discrete FEM space with respect to a norm that is stronger than $\|\cdot\|_{1,\varepsilon}$. This indicates that it is a more stable method than the Galerkin FEM which by (6.12) is coercive only with respect to $\|\cdot\|_{1,\varepsilon}$.

Remark 6.16. When S^N comprises piecewise linears or bilinears (i.e., $k=1$), the term Δw^N is zero and disappears from the proof of Lemma 6.15; consequently, the condition (6.22) can be relaxed to $2\delta_\tau \le C_5/\|b\|^2_{L_\infty(\tau)}$. Furthermore, one no longer needs the inverse inequality (6.20) so Lemma 6.15 is valid on any mesh, including a Shishkin mesh.

One can exploit Lemma 6.15 to derive an error estimate in a fairly standard way. Set $h := \max_\tau h_\tau$, the mesh diameter.

Lemma 6.17. *Choose δ_τ such that (6.22) is satisfied. Then*

$$\|u^I - u_{SD}\|_{SD} \le Ch^k \left[\sum_{\tau\in\Omega^N} \left(\varepsilon + \delta_\tau + \delta_\tau^{-1} h_\tau^2 + h_\tau^2\right) |u|_{k+1,\tau}^2 \right]^{1/2} . \tag{6.24}$$

Proof. By Lemma 6.15 we have

$$\|u^I - u_{SD}\|_{SD}^2 \le 2\, B_{SD}(u^I - u_{SD}, u^I - u_{SD}) = 2\, B_{SD}(u^I - u, u^I - u_{SD}),$$

using the Galerkin orthogonality property (6.18). Now we estimate the right-hand side term by term, invoking the Cauchy–Schwarz inequality and the interpolation error estimates (6.19):

$$\begin{aligned} \left|\varepsilon(\nabla(u^I - u), \nabla(u^I - u_{SD}))\right| &\le \varepsilon^{1/2} |u^I - u|_1 \, \|u^I - u_{SD}\|_{SD} \\ &\le C\varepsilon^{1/2} h^k |u|_{k+1} \, \|u^I - u_{SD}\|_{SD} ; \end{aligned}$$

$$\begin{aligned} &\left|(\mathbf{a}\cdot\nabla(u^I - u) + b(u^I - u), u^I - u_{SD})\right| \\ &\quad = \left((b - \nabla\cdot\mathbf{a})(u^I - u), u^I - u_{SD}\right) - \left(u^I - u, \mathbf{a}\cdot\nabla(u^I - u_{SD})\right) \\ &\quad \le C \left[\left(\sum_{\tau\in\Omega^N} \|u^I - u\|_{0,\tau}^2 \right)^{1/2} + \left(\sum_{\tau\in\Omega^N} \delta_\tau^{-1} \|u^I - u\|_{0,\tau}^2 \right)^{1/2} \right] \|u^I - u_{SD}\|_{SD} \\ &\quad \le C h^k \left[\sum_{\tau\in\Omega^N} h_\tau^2 (1 + \delta_\tau^{-1}) |u|_{k+1,\tau}^2 \right]^{1/2} \|u^I - u_{SD}\|_{SD} ; \end{aligned}$$

and

$$\begin{aligned} &\left| \sum_{\tau\in\Omega^N} \delta_\tau \big(-\varepsilon\Delta(u^I - u) + \mathbf{a}\cdot\nabla(u^I - u) + b(u^I - u), \mathbf{a}\cdot\nabla(u^I - u_{SD}) \big)_\tau \right| \\ &\quad \le C \sum_{\tau\in\Omega^N} \delta_\tau^{1/2} (\varepsilon h_\tau^{k-1} + h_\tau^k + h_\tau^{k+1}) |u|_{k+1,\tau} \, \delta_\tau^{1/2} \|\mathbf{a}\cdot\nabla(u^I - u_{SD})\|_{0,\tau} \\ &\quad \le C \left[\sum_{\tau\in\Omega^N} (\varepsilon + \delta_\tau) h_\tau^{2k} |u|_{k+1,\tau}^2 \right]^{1/2} \|u^I - u_{SD}\|_{SD} . \end{aligned}$$

In the last inequality we used the bound $\varepsilon\delta_\tau \le Ch_\tau^2$, which is implied by (6.22). Adding all of these estimates, we obtain (6.24). □

In order to extract the best possible rate of convergence from Lemma 6.17 while honouring the constraint (6.22) on δ_τ, we balance the various terms in (6.24) by choosing

$$(6.25)\qquad \delta_\tau = \begin{cases} \delta_0 h_\tau & \text{for } Pe_\tau > 1, \\ \delta_1 h_\tau^2/\varepsilon & \text{for } Pe_\tau \le 1, \end{cases}$$

where the *mesh Péclet number* is defined to be $Pe_\tau := \|\mathbf{a}\|_{L^\infty(\tau)} h_\tau/\varepsilon$. Here δ_0 and δ_1 are user-chosen positive constants. The more important case $Pe_\tau > 1$ is usually referred to as the convection-dominated case.

Exercise 6.18. Verify that the choice of δ_τ in (6.25) does yield the best possible rate of convergence in Lemma 6.17 subject to the constraint (6.22).

Theorem 6.19. *Choose δ_T such that* (6.22) *and* (6.25) *are satisfied. Then the solution u_{SD} computed by the SDFEM satisfies the global error estimate*

$$\|u - u_{SD}\|_{SD} \le C\,(\varepsilon^{1/2} + h^{1/2})\, h^k |u|_{k+1}\,.$$

Proof. Substituting the choice (6.25) into (6.24) yields

$$\|u^I - u_{SD}\|_{SD} \le C\,(\varepsilon^{1/2} + h^{1/2})\, h^k |u|_{k+1}.$$

One can verify readily that (6.19) implies

$$\|u - u^I\|_{SD} \le C\,(\varepsilon^{1/2} + h^{1/2})\, h^k |u|_{k+1}.$$

The theorem now follows from a triangle inequality. □

Remark 6.20 (Optimality of the error in various norms)**.** In the convection-dominated case we have $\varepsilon < \|\mathbf{a}\|_{L_\infty(\tau)}\, h_\tau/2$ and hence Theorem 6.19 gives the global estimate

$$(6.26)\qquad \|u - u_{SD}\|_0 + \Big(\sum_{\tau\in\Omega^N} \delta_\tau \|\mathbf{a}\cdot\nabla(u - u_{SD})\|_{0,\tau}^2\Big)^{1/2} \le C h^{k+1/2} |u|_{k+1},$$

with $\delta_\tau = \delta_0 h_\tau$. For comparison, (6.19) informs us that

$$\|u - u^I\|_0 \le C h^{k+1} |u|_{k+1} \quad\text{and}\quad |u - u^I|_1 \le C h^k |u|_{k+1}.$$

Thus we see that the L_2 error of the computed derivative in the streamline direction is optimal, but the L_2 error of the computed solution is order 1/2 less than optimal.

This apparent loss of accuracy in the L_2 norm has attracted much attention. Zhou [**Zho97**] constructed a simple example for piecewise linears on a special quasi-uniform mesh where the SDFEM converged with order only 1.5, but it is not known whether a similar loss of optimal-order accuracy in L_2 can occur for other choices of the FEM trial space S^N. Other stabilized FEMs for convection-diffusion problems suffer the same gap between theory and practice; see [**RS15c**].

In particular cases some optimal results are known. Optimal convergence in $L_2(\Omega)$ for linears and bilinears on a Shishkin mesh on $\Omega = (0,1)^2$ is proved in [**ST03, ZLY16**]; see Theorem 6.48 and Remark 6.49 below. In [**CX08**] Chen and Xu prove quasi-optimality in the L_∞ norm on an arbitrary grid for the solution of a modified SDFEM for a one-dimensional convection-diffusion problem. In a very technical paper, Sangalli [**San03**] shows that in the one-dimensional case (2.14), on an equidistant grid the SDFEM yields a solution that is quasi-optimal with respect to a certain interpolated norm that is roughly similar to our norm $\|\cdot\|_{SD}$.

Remark 6.21 (Optimal choice of δ_τ)**.** No precise general formula for an "optimal" (in some sense) value of the SDFEM parameter δ_τ is known; the choice (6.25) seems to be the best statement that one can make. There has been much research into this question: for discussions of how to choose δ_τ see, e.g., [**AT04, BR94, FRSW99, HS01, MS96, RST08**] and more recently, [**JKS11, JN13**].

In (6.26) the term $|u|_{k+1}$ is typically $\mathcal{O}\left(\varepsilon^{-k-1/2}\right)$. In general this will dominate the $h^{k+1/2}$ term and consequently (6.26) does not imply that the error $u - u_{SD}$ is small in some norm. Thus this estimate is of limited value. Nevertheless, one can choose some maximal subset $\hat{\Omega}$ of Ω that excludes all layers, restrict the norms in (6.26) to $\hat{\Omega}$, then prove essentially the same bounds again (in terms of the new norms) by means of cut-off functions [**RST08**, Section III.3.2.1].

Exercise 6.22. Take $\delta_\tau = 0$ for each $\tau \in \Omega^N$, so the SDFEM becomes the standard Galerkin method (which is of course unstable in general). Imitate the above analysis of the SDFEM to obtain

$$\|u - u_{\mathrm{Gal}}\|_{1,\varepsilon} \le Ch^k |u|_{k+1}, \tag{6.27}$$

where u_{Gal} is the solution computed by the Galerkin method.

Recall that typically $|u|_{k+1} = \mathcal{O}\left(\varepsilon^{-k+1/2}\right)$ because of exponential boundary layers, so the bound (6.27) does not imply that the error in the Galerkin solution is small. Nevertheless, bounds like this are sometimes presented imprecisely as $\|u - u_{\mathrm{Gal}}\|_{1,\varepsilon} \le Ch^k$; i.e., the dependence on $|u|_{k+1}$ is hidden—then it is used to assert misleadingly that the method will yield an accurate computed solution! Thus in error bounds for the numerical solutions of singularly perturbed problems, one must always examine carefully the part played by norms of the true solution.

There are two crucial differences between Theorem 6.19 and (6.27): the norm $\|\cdot\|_{1,\varepsilon}$ is too weak to suppress large oscillations in the computed solution, and (unlike Theorem 6.19) a result like (6.27) cannot be localised away from layers using cut-off functions because these large oscillations spread far from the boundary layers.

Remark 6.23 (Stability of the SDFEM in different directions)**.** Lemma 6.15 and (6.17) together imply an a priori estimate for the SDFEM solution u_{SD},

$$\|u_{SD}\|_{SD} \le C\Big(\|f\|_0^2 + \sum_{\tau} \delta_\tau \|f\|_{0,T}^2\Big)^{1/2}. \tag{6.28}$$

Thus the method imposes some control on the streamline derivative $\mathbf{a}\cdot\nabla u_{SD}$ of the computed solution. In the more interesting convection-dominated case, with $\delta_\tau = \delta_0 h_\tau$, inequality (6.28) says essentially that $\|\mathbf{a}\cdot\nabla u_{SD}\|_{0,\tau}$ can be at most $\mathcal{O}\left(h_\tau^{1/2}\right)$. This property distinguishes the SDFEM from a standard Galerkin method, for whose oscillatory solution u^N one can have $\|\mathbf{a}\cdot\nabla u^N\|_{0,\tau} = \mathcal{O}(1)$.

This enhanced stability in the subcharacteristic direction means that the SDFEM can compute fairly satisfactory exponential boundary layers in solutions of convection-diffusion problems, provided that δ_τ is chosen carefully. Note however that the SDFEM contains no mechanism for stabilization perpendicular to the subcharacteristics and, consequently, along characteristic layers the computed solution typically displays oscillations; as usual with such layers, these oscillations are confined to a fairly small neighbourhood of the layer.

Kopteva [**Kop04**] gives a detailed analysis of the accuracy of the SDFEM inside characteristic (boundary and interior) layers.

In [**MS96**] the authors investigate numerically the effect that changing the value of δ_τ (the same value is used for all τ for simplicity) has on the computed solution for a problem with exponential outflow boundary layers and a characteristic interior layer. Comparing [**MS96**, Figures 2 and 3], one sees that changing δ_τ can affect significantly the computed solution in the outflow layers but has little effect on the computed interior layer, which exhibits localised oscillations; this behaviour is consistent with the observations made in Remark 6.23.

Remark 6.24. In order to reduce or remove any oscillations that appear along characteristic layers, several authors have modified the SDFEM by adding artificial crosswind diffusion to the PDE or even by introducing nonlinear "shock-capturing" terms into the SDFEM formulation. See for example [**Cod11, KLR02, LR06, SE00**].

For further analysis of the SDFEM, see [**RST08**].

6.5. Stability of the Galerkin FEM for higher-degree polynomials

Earlier in this chapter we dismissed the (Bubnov–)Galerkin FEM as an unstable method that is unsuited for the solution of convection-diffusion problems on standard meshes. However, the stability or instability of an FEM can be a subtle affair. In the present section we show that when piecewise polynomials *of sufficiently high degree* k are used as the finite element basis functions, then the Galerkin FEM solution is more stable than the previous analysis suggests, and the poor behaviour of the method is caused only by a certain component of the computed solution. Consequently, one needs to stabilize only that component (i.e., not all of the computed solution) in order to obtain a viable method.

Our presentation is based on the elegant paper of Knobloch and Tobiska [**KT11**].

Consider a shape-regular family of meshes $\{\Omega^N\}$, where each Ω^N is a partition of the bounded domain $\Omega \subset \mathbb{R}^2$ using triangles. Once again we use the notation of Remark 6.14. The finite element space $S^N \subset H_0^1(\Omega)$ comprises piecewise polynomials of degree $k \geq 3$ defined on the triangular mesh Ω^N, i.e., $S^N\big|_\tau = P_k(\tau)$ for each triangle $\tau \in \Omega^N$. To solve (6.8) numerically, we use the standard Galerkin bilinear form $B(\cdot,,\cdot)$ of (6.10). That is, our finite element solution $u_{\text{Gal}} \in S^N$ is defined by

$$B(u_{\text{Gal}}, w^N) = (f, w^N) \quad \text{for all } w^N \in S^N.$$

The coercivity inequality (6.12) implies that u_{Gal} is well-defined.

The shape-regularity of the mesh implies that (cf. (6.20)) one has the inverse inequality

$$|w^N| \leq C_{\text{inv}} h_\tau^{-1} \|w^N\|_{0,\tau} \quad \forall \tau \in \Omega^N, \tag{6.29}$$

where the positive constant C_{inv} is independent of τ.

A crucial ingredient in our analysis is the following. Assume that for each $\tau \in \Omega^N$, the restriction of our finite element space S^N to τ contains a nontrivial "bubble space" $\mathcal{A}(\tau)$. That is, for each τ there is some function in $S^N|_\tau \cap H_0^1(\tau) =: \mathcal{A}(\tau)$ that is not identically zero. These functions are called bubble functions because of their appearance: they vanish on the boundary of τ.

Figure 6.2 shows a one-dimensional bubble function defined on the interval $[0, h]$, and Figure 6.3 gives two distinct views of a two-dimensional bubble function defined on a triangle in the (x, y)-plane with vertices $(0, 0)$, $(h, 0)$, $(0, h)$. (From the latter figure you can see why they are called bubbles.)

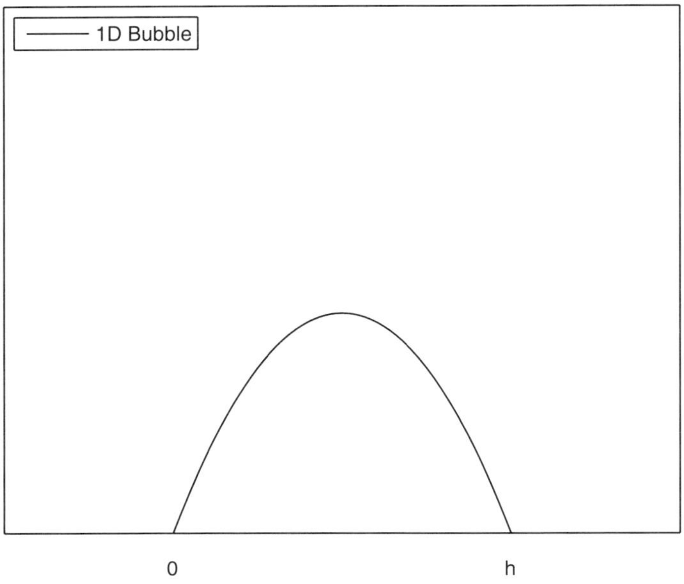

Figure 6.2. Bubble function in one dimension

With standard piecewise polynomial trial spaces, it is easy to see that in one dimension one needs at least quadratics to obtain bubbles, while in two dimensions one needs polynomials of degree at least 3.

Remark 6.25. On each $\tau \in \Omega^N$ with $\Omega \subset \mathbb{R}^2$, the bubble subspace

$$\mathcal{A}(\tau) = \mathcal{B}\, P_{k-3}(\tau) := \{\mathcal{B}v : v \in P_{k-3}(\tau)\},$$

where $\mathcal{B}$ is the product of the barycentric coordinates of τ.

Exercise 6.26. Prove Remark 6.25.

Exercise 6.27. In the one-dimensional case, where τ is a (bounded) interval, determine and prove the analogue of Remark 6.25.

Now

$$\mathcal{A}^N := \bigoplus_{\tau \in \Omega^N} \mathcal{A}(\tau)$$

is a nonempty subspace of S^N.

To prove stability of a bilinear form with respect to a norm, the simplest way is to use coercivity (as in Lemma 6.15), but coercivity does not always hold true, and then one must instead use the more general framework (see Exercise 6.29) of an inf-sup condition.

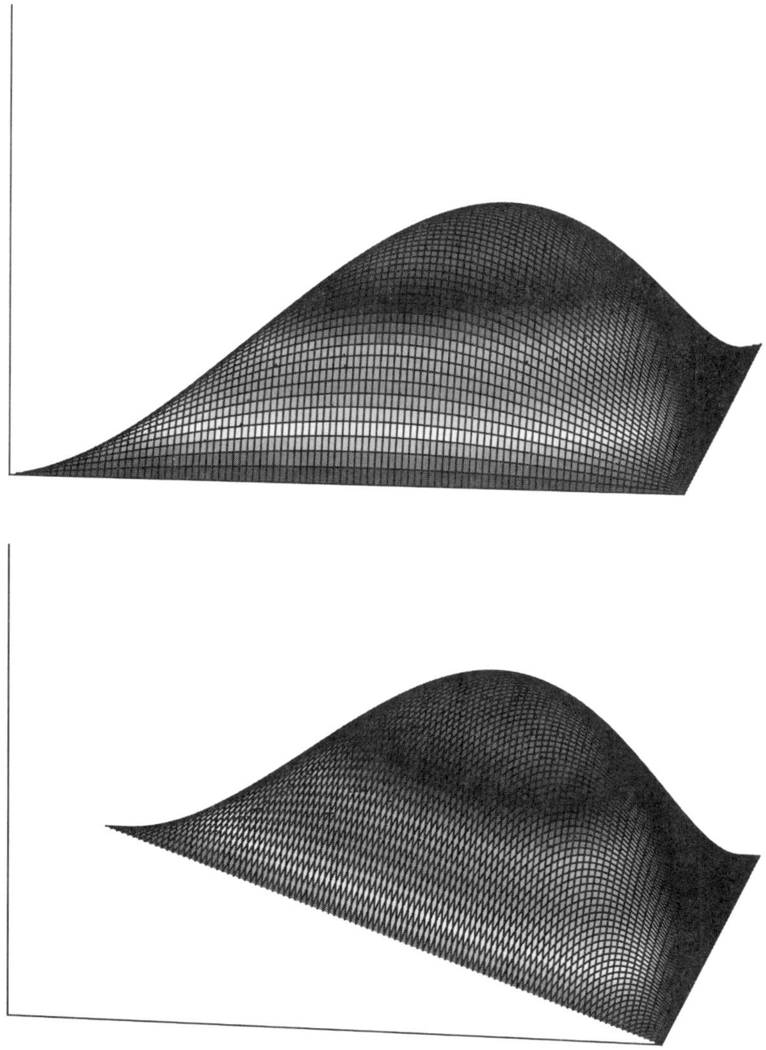

Figure 6.3. Two views of a bubble function in two dimensions

Recall the streamline diffusion norm $\|\cdot\|_{SD}$ that was defined in (6.21). Define a *weaker variant* $\|\cdot\|_{SDw}$ of this norm:

$$\|v\|_{SDw} := \Big(\varepsilon|v|_1^2 + \sum_{\tau\in\Omega^N} \delta_\tau \|\Pi_\tau(\mathbf{a}\cdot\nabla v)\|_{0,\tau}^2 + C_5\|v\|_0^2\Big)^{1/2}, \tag{6.30}$$

where Π_τ is the orthogonal $L^2(\tau)$ projection of $L^2(\tau)$ onto $\mathcal{A}(\tau)$. It is easy to see that $\|v\|_{SDw} \le \|v\|_{SD}$ for all v for which these norms are defined. Throughout this section, the quantity δ_τ in (6.30) is as specified in the next theorem.

Theorem 6.28. *Assume that in* (6.21) *one has*

$$0 \le \delta_\tau \le \frac{C_6 h_\tau^2}{\max\{\varepsilon, h_\tau \|\mathbf{a}\|_{0,\infty,\tau}\}} \quad \text{for all } \tau \in \Omega^N, \tag{6.31}$$

where the constant C_6 *is independent of* h *and* ε. *Then there exists a positive constant* $\tilde{C}$, *which is independent of* h *and* ε, *such that*

$$\inf_{w^N \in S^N} \sup_{v^N \in S^N} \frac{B(w^N, v^N)}{\|w^N\|_{SDw} \|v^N\|_{SDw}} \ge \tilde{C}.$$

Proof. By (6.12) we have

$$B(w^N, w^N) \ge \min\{1, C_5\} \|w^N\|_{1,\varepsilon}^2 \quad \forall w^N \in S^N.$$

This does not prove the theorem as the term $\sum_{\tau \in \Omega^N} \delta_\tau \|\Pi_\tau(\mathbf{a} \cdot \nabla v)\|_{0,\tau}^2$ is missing from the right-hand side. Thus, given any $w^N \in S^N$, we shall construct a function $v^N \in S^N$ such that

$$B(w^N, v^N) \ge \|w^N\|_{SDw}^2 \quad \text{and} \quad \|w^N\|_{SDw} \ge \tilde{C} \|v^N\|_{SDw}. \tag{6.32}$$

It is clear that the theorem will follow from this pair of inequalities.

First, define a bubble function $z^N \in \mathcal{A}^N$ by

$$z^N\big|_\tau := \delta_\tau \Pi_\tau(\mathbf{a} \cdot \nabla w^N) \quad \forall \tau \in \Omega^N.$$

Then

$$(\mathbf{a} \cdot \nabla w^N, z^N)_\tau = \delta_\tau \|\Pi_\tau(\mathbf{a} \cdot \nabla w^N)\|_{0,\tau}^2 \quad \forall \tau \in \Omega^N.$$

Hence

$$B(w^N, z^N) = \sum_{\tau \in \Omega^N} \delta_\tau \|\Pi_\tau(\mathbf{a} \cdot \nabla w^N)\|_{0,\tau}^2 + \varepsilon(\nabla w^N, \nabla z^N) + (c w^N, z^N). \tag{6.33}$$

Now the inverse inequality (6.29) and the assumption (6.31) yield

$$\varepsilon \big|z^N\big|_{1,\tau}^2 \le C_{\text{inv}}^2 \varepsilon h_\tau^{-2} \|z^N\|_{0,\tau}^2 \le C_1 C_{\text{inv}}^2 \delta_\tau \|\Pi_\tau(\mathbf{a} \cdot \nabla w^N)\|_{0,\tau}^2$$

and

$$\|z^N\|_{0,\tau} \le \delta_\tau \|\mathbf{a}\|_{0,\infty,\tau} |w^N|_{1,\tau} \le C_1 C_{\text{inv}} \|w^N\|_{0,\tau}. \tag{6.34}$$

Thus, applying the inequality $st \le \sigma s^2/2 + t^2/(2\sigma)$ for all $\sigma > 0$ and (6.9), we get

$$\begin{aligned} &\big|\varepsilon(\nabla w^N, \nabla z^N)_\tau + (c w^N, z^N)_\tau\big| \\ &\quad \le \varepsilon |w^N|_{1,\tau} |z^N|_{1,\tau} + \|c\|_{0,\infty,\tau} \|w^N\|_{0,\tau} \|z^N\|_{0,\tau} \\ &\quad \le \hat{C}\big[\varepsilon |w^N|_{1,\tau}^2 + C_5 \|w^N\|_{0,\tau}^2\big] + \frac{\delta_\tau}{2} \|\Pi_\tau(\mathbf{a} \cdot \nabla w^N)\|_{0,\tau}^2 \end{aligned}$$

for some constant $\hat{C}$. Sum this inequality over all $\tau \in \Omega^N$, then apply it to (6.33), obtaining

$$B(w^N, z^N) \geq \frac{1}{2} \sum_{\tau\in\Omega^N} \delta_\tau \|\Pi_\tau(\mathbf{a}\cdot\nabla w^N)\|_{0,\tau}^2 - \hat{C}B(w^N, w^N),$$

where we also used the coercivity inequality (6.12). Setting

$$v^N := 2z^N + (1 + 2\hat{C})w^N,$$

the previous inequality yields the first inequality in (6.32).

To establish the second inequality in (6.32), note first that the inverse inequality (6.29), the assumption (6.31) on δ_τ, and the definition of z^N imply that for each τ one has

$$|z^N|_{1,\tau} \leq C_{\text{inv}} h_\tau^{-1}\delta_\tau \|\mathbf{a}\|_{0,\infty,\tau}|w^N|_{1,\tau} \leq C_{\text{inv}}C_6|w^N|_{1,\tau}$$

and

$$\|\Pi_\tau(\mathbf{a}\cdot\nabla z^N)\|_{0,\tau} \leq \|\mathbf{a}\|_{0,\infty,\tau}|z^N|_{1,\tau} \leq C_{\text{inv}}C_6\|\Pi_\tau(\mathbf{a}\cdot\nabla w^N)\|_{0,\tau}.$$

These inequalities and (6.34) yield the second inequality in (6.32). □

Exercise 6.29. Suppose that for some norm $|||\cdot|||$ one has the coercivity property $B(v^N, v^N) \geq C|||v^N|||^2$ for all $v^N \in S^N$, with some constant C. Show that this implies the inf-sup condition

$$\inf_{w^N\in S^N} \sup_{v^N\in S^N} \frac{B(w^N, v^N)}{|||w^N|||\cdot|||v^N|||} \geq C.$$

For simplicity,

we assume that **a** *is constant in the remainder of section* 6.5.

The general case of variable **a** is discussed in [**KT11**].

Set

$$\hat{S}^N = \{v^N \in H^1(\Omega) : v^N|_\tau \in P_{k-2}(\tau)\, \forall \tau \in \Omega^N\}.$$

Clearly, $\hat{S}^N \cap H_0^1(\Omega) \subsetneq S^N$.

Lemma 6.30. *The norms* $\|\cdot\|_{SD}$ *and* $\|\cdot\|_{SDw}$ *are equivalent on* $\hat{S}^N$. *(Here the same values of the* δ_τ *are used in each norm.)*

Proof. For all $v^N \in \hat{S}^N$ one clearly has $\|v^N\|_{SDw} \leq \|v^N\|_{SD}$, so we need show only that $\|v^N\|_{SDw} \geq C\|v^N\|_{SD}$ for some positive constant C that is independent of v^N. This will be done by considering separately each $\tau \in \Omega^N$.

Fix $\tau \in \Omega^N$. Then $\hat{S}^N\big|_\tau = P_{k-2}(\tau)$. Recall from Remark 6.25 that $\mathcal{A}(\tau) = \mathcal{B}P_{k-3}(\tau)$. Let $\mathcal{A}(\tau)^\perp$ denote the orthogonal complement of $\mathcal{A}(\tau)$ in $L^2(\tau)$. A key observation is that $\mathcal{A}(\tau)^\perp \cap P_{k-3}(\tau) = \{0\}$: for if $w \in P_{k-3}(\tau)$, then $(w)(\mathcal{B}w) = 0$ is possible only if $w = 0$.

We introduce another projection. Let π_τ be the orthogonal $L^2(\tau)$ projection of $L^2(\tau)$ onto $P_{k-3}(\tau)$. If $z \in L^2(\tau)$ and $\Pi_\tau z = 0 = (I - \pi_\tau)z$, where I is the identity mapping, this says that $z \in \mathcal{A}(\tau)^\perp$ and $z \in P_{k-3}(\tau)$, so by our earlier observation one must have $z = 0$. Consequently,

$$z \mapsto \left[\|\Pi_\tau(z)\|_{0,\tau}^2 + \|(I - \pi_\tau)(z)\|_{0,\tau}^2\right]^{1/2}$$

is a norm on $L^2(\tau)$, and a fortiori a norm on $P_{k-1}(\tau)$. As all norms on a finite-dimensional space are equivalent, it follows that there exists a positive constant C_7 such that

$$(6.35)\quad \|\Pi_\tau(\mathbf{a}\cdot\nabla v)\|_{0,\tau}^2 + \|(I - \pi_\tau)(\mathbf{a}\cdot\nabla v)\|_{0,\tau}^2 \geq C_7\|\mathbf{a}\cdot\nabla v\|_{0,\tau}^2 \quad \forall v \in P_k(\tau).$$

Here, to verify that the constant C_7 is independent of v and τ, one should transform from τ to a reference triangle of unit size, apply the norm equivalence, then transform back to τ. The details of these transformations are in [**KT11**].

Finally, if $y \in P_{k-2}(\tau)$, then $\mathbf{a}\cdot\nabla y \in P_{k-3}(\tau)$, so $(I-\pi_\tau)(\mathbf{a}\cdot\nabla y) = 0$ and (6.35) yields $\|\Pi_\tau(\mathbf{a}\cdot\nabla y)\|_{0,\tau}^2 \geq C_7\|\mathbf{a}\cdot\nabla y\|_{0,\tau}^2$. This is the desired inequality, and we are done. □

Lemma 6.30 tells us that for discretisations of the two-dimensional problem (6.8) on shape-regular triangular meshes with globally continuous piecewise polynomials P_k of degree $k \geq 3$, the standard Galerkin FEM is just as stable as the streamline diffusion FEM on a significant subspace P_{k-2} of P_k. Thus the instability of the Galerkin method is due to the complement of P_{k-2} in P_k.

Remark 6.31 (The one-dimensional case)**.** Consider the one-dimensional situation, i.e., the problem (6.1) on an arbitrary mesh (in one dimension all meshes are shape-regular), where we use the Galerkin FEM with piecewise polynomials of degree $k \geq 2$. Retracing our earlier arguments, one finds now that (cf. Exercise 6.27) on each mesh interval τ the bubble subspace $\mathcal{A}(\tau) = \mathcal{B}\,P_{k-2}(\tau)$; note the change from the two-dimensional case. Likewise, in Lemma 6.30, one now has

$$\hat{S}^N = \{v^N \in H^1(\Omega) : v^N|_\tau \in P_{k-1}(\tau)\,\forall\tau \in \Omega^N\}.$$

Thus in the Galerkin solution, the subspace of piecewise polynomials of degree $k-1$ is controlled by the streamline diffusion norm and so is stable; only the polynomials of highest degree k can cause instabilities.

Let's take $k = 2$ and discuss in detail what happens. Write the mesh as $x_0 < x_1 < \cdots < x_N$. Then for each mesh interval $\tau = [x_{i-1}, x_i]$, the operator Π_τ is the orthogonal L^2 projection of $L^2(\tau)$ onto the one-dimensional space spanned by the bubble function $\mathcal{B}_i(x) := (x - x_{i-1})(x - x_i)$. But

$$\int_{x_{i-1}}^{x_i} \mathcal{B}_i'(x)\mathcal{B}_i(x)\,dx = \int_{x_{i-1}}^{x_i} \frac{1}{2}\left(\mathcal{B}_i(x)^2\right)'\,dx = \frac{1}{2}\left[\mathcal{B}_i(x_i)^2 - \mathcal{B}_i(x_{i-1})^2\right] = 0,$$

so by its definition $\Pi_\tau(\mathbf{a}\cdot\nabla\mathcal{B}_i) = a\,\Pi_\tau\mathcal{B}_i' = 0$. Consequently, the component $\sum_{\tau\in\Omega^N}\delta_\tau\|\Pi_\tau(\mathbf{a}\cdot\nabla v)\|_{0,\tau}^2$ of the norm $\|\cdot\|_{SDw}$ exerts no control over multiples of $\mathcal{B}_i$ in the computed solution. That is, the computed Galerkin solution can—and generally does—exhibit fast oscillations.

As we now know that the misbehaviour of computed Galerkin solutions when solving (6.8) can be caused only by a particular component of the finite element space, it follows that only that component needs to be stabilized. One way of doing this is to employ *local projection stabilization*, which is the main topic of [**KT11**]. This FEM technique for convection-dominated problems is an alternative to streamline diffusion stabilization which has been investigated in many papers; we do not discuss it here but instead refer the reader to [**KT11**, **Roo12**, **RST08**].

Remark 6.32. A different point of view is taken in [**Sch08**]; although this paper discusses a stabilized FEM, [**Sch08**, Remark 5.2] points out that the main theoretical result can be restricted to the Galerkin FEM. If the solution of the reduced problem associated with (6.8) is smooth, it is shown that an ε-weighted weak imposition of the boundary conditions (cf. Example 6.52 below) can yield greatly improved numerical results.

6.6. Shishkin meshes

FEMs can of course be implemented on Shishkin meshes (which we denote by Ω_S^N) like those of Figures 5.2 and 5.3; the mesh rectangles can be bisected by one of their diagonals into triangles to permit the use of, e.g., a piecewise linear FEM, though the way in which this bisection is done can in some cases influence strongly the computed solution [**Kop14**].

Throughout section 6.6 we consider the convection-diffusion problem

$$-\varepsilon\Delta u(x,y) + \mathbf{a}(x,y)\cdot\nabla u(x,y) + b(x,y)u(x,y) = f(x,y) \quad \text{on } \Omega = (0,1)^2,$$
$$u(x,y) = 0 \quad \text{on } \partial\Omega,$$

with $\mathbf{a} = (a_1, a_2) \geq (\alpha_1, \alpha_2) > (0,0)$. Its solution u exhibits two exponential outflow layers along the sides $x = 1$ and $y = 1$ of Ω, as in Example 4.5. We shall assume the decomposition (4.14) of u.

To describe the analysis, we label the different regions of the Shishkin mesh of Figure 5.2 as in Figure 6.4. Furthermore, for the mesh transition points we take

$$\lambda_x = \frac{m\varepsilon}{\alpha_1}\ln N, \quad \lambda_y = \frac{m\varepsilon}{\alpha_2}\ln N,$$

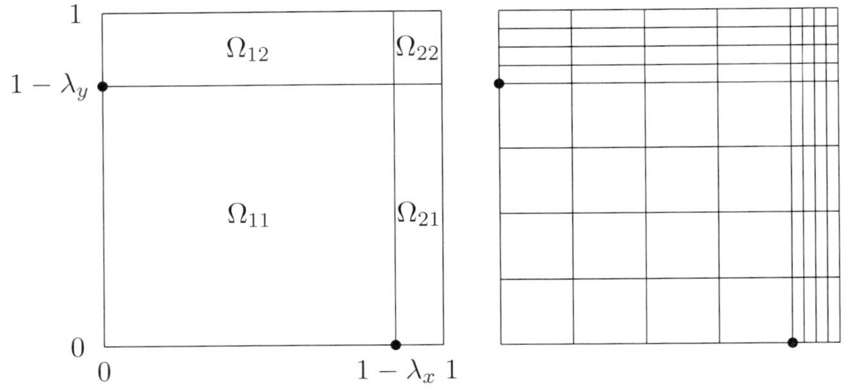

Figure 6.4. Regions in a Shishkin mesh Ω_S^N with $N = 8$ for two exponential outflow layers

where $m = 2$ in section 6.6.1 and $m = 5/2$ in section 6.6.2. (One generally chooses for m the smallest value that enables the analysis to work; thus, to justify the above two choices, one must go through all the details of the analysis in these two subsections.)

In the analysis that follows, we remind the reader that C is a generic positive constant that is independent of the mesh.

In Figure 6.4, note that the mesh rectangles in $\Omega_{12} \cup \Omega_{21}$ have a high aspect ratio, i.e., their length greatly exceeds their width. To analyse such methods, the highly anisotropic nature of the mesh necessitates the use of sharp anisotropic interpolation estimates for general meshes, which we now describe.

Lemma 6.33. *Suppose that each element τ (triangle or rectangle) of a mesh is contained in a rectangle with side lengths (h_x, h_y) and contains a rectangle with side lengths (Ch_x, Ch_y). In the case of triangles, assume also a maximum angle condition: the interior angles are bounded away from π. Let $v \in H^2(\tau)$. Let v^I denote the nodal interpolant (linear or bilinear) of v. Write $\|\cdot\|_{0,\tau}$ for the norm in $L_2(\tau)$. Then* [**AD92**, **Ape99**]

$$\|v - v^I\|_{0,\tau}^2 \le C \sum_{|\alpha|=2} h^{2\alpha} \|D^\alpha v\|_{0,\tau}^2, \tag{6.36a}$$

$$\|\partial_x(v - v^I)\|_{0,\tau}^2 \le C \sum_{|\alpha|=1} h^{2\alpha} \|D^\alpha \partial_x v\|_{0,\tau}^2, \tag{6.36b}$$

$$\|\partial_y(v - v^I)\|_{0,\tau}^2 \le C \sum_{|\alpha|=1} h^{2\alpha} \|D^\alpha \partial_y v\|_{0,\tau}^2. \tag{6.36c}$$

Here α is the multi-index (α_1, α_2), $|\alpha| = \alpha_1 + \alpha_2$, $h^\alpha = h_x^{\alpha_1} h_y^{\alpha_2}$, and

$$D^\alpha = \frac{\partial^{\alpha_1}}{\partial x^{\alpha_1}} \frac{\partial^{\alpha_2}}{\partial y^{\alpha_2}}.$$

The rectangular Shishkin mesh of Figure 6.4 and the triangular Shishkin mesh obtained by bisecting each mesh rectangle will satisfy the hypotheses of Lemma 6.33. The anisotropic estimates (6.36) are valid on any mesh satisfying these hypotheses; bounds of this type are useful on Shishkin meshes because of the very small mesh width factor in precisely the coordinate direction where the solution derivative is large. Standard isotropic interpolation error estimates use only the diameter of the element and thereby lose the benefit of the Shishkin mesh in the regions Ω_{12} and Ω_{21} of Figure 6.4 when analysing the interpolation error.

Using Lemma 6.33 and the decomposition (4.14) of u, one can show [**DR97**] (or see [**RST08**, Section III.3.5.2]) that for piecewise linear or bilinear interpolation on a Shishkin mesh, one has

$$\|u - u^I\|_{L_\infty(\Omega)} \le C(N^{-1}\ln N)^2, \tag{6.37a}$$

$$\|u - u^I\|_{1,\varepsilon} \le CN^{-1}\ln N, \tag{6.37b}$$

and

$$\|u - u^I\|_0 \le CN^{-2} + C\sqrt{\varepsilon}(N^{-1}\ln N)^2, \tag{6.37c}$$

so

$$\|u - u^I\|_0 \le CN^{-2} \quad \text{when } \sqrt{\varepsilon} \le C(\ln N)^{-2}. \tag{6.37d}$$

These sharp bounds tell us the convergence rates we can hope for when devising FEMs for convection-diffusion problems on Shishkin meshes.

Exercise 6.34. Use Lemma 6.33 and the decomposition (4.14) to prove the interpolation error estimates (6.37).

The next exercise demonstrates how to get interpolation error estimates using only bounds on the derivatives of the true solution—no decomposition of that solution is used.

Exercise 6.35 (Based on [**Lin01**]). Consider the one-dimensional convection-diffusion problem (2.14) for which Theorem 2.27 gives bounds on the derivatives of the true solution u. Let u^I denote the piecewise linear interpolant to u on the Shishkin mesh $0 = x_0 < x_1 < \cdots < x_N = 1$ of Section 3.4.

(i) For $x \in [x_{i-1}, x_i]$ and $i = 1, 2, \ldots, N$, show that

$$(u - u^I)(x) = \int_{x_{i-1}}^{x} (x - s)u''(s)\,ds - \frac{x - x_{i-1}}{x_i - x_{i-1}} \int_{x_{i-1}}^{x_i} (x_i - s)u''(s)\,ds.$$

(ii) Suppose the function g is positive and monotonically increasing on $[x_{i-1}, x_i]$. Prove that

$$\int_{x_{i-1}}^{y} (y-s)g(s)\,ds \le \frac{1}{2}\left[\int_{x_{i-1}}^{y} \sqrt{g(s)}\,ds\right]^2 \quad \text{for } y \in [x_{i-1}, x_i].$$

Hint. Consider both sides of the inequality as functions of y.

(iii) Combine (i) and (ii) to get

$$\|u - u^I\|_{L^\infty[0,1]} \le C(N^{-1}\ln N)^2$$

for some constant C.

(iv) Deduce from (iii) that $\|u - u^I\|_{1,\varepsilon} \le CN^{-1}\ln N$ for some constant C, using $\|u - u^I\|_{L^2[0,1]} \le \|u - u^I\|_{L^\infty[0,1]}$ and

$$\int_0^1 \left[(u-u^I)'(x)\right]^2 dx = -\int_0^1 (u-u^I)(x)u''(x)\,dx \le C\varepsilon^{-1}\|u - u^I\|_{L^\infty[0,1]}$$

by Theorem 2.27.

6.6.1. Galerkin FEM with bilinears. To see in detail how a finite element analysis for a two-dimensional problem is carried out on a Shishkin mesh, we consider here the simplest situation: the standard Galerkin FEM, based on the bilinear form $B(\cdot,\cdot)$ of (6.10), applied to the problem described at the start of section 6.6. Thus the solution has two exponential boundary layers and can be decomposed according to (4.14). This analysis is based on [**SO97**].

Since $u \equiv 0$ on $\partial\Omega$, we have $u \in H_0^1(\Omega)$. Let $S^N \subset H_0^1(\Omega)$ be the space of piecewise bilinear functions on the Shishkin mesh Ω_S^N that vanishes on the boundary $\partial\Omega$. Let u_{Gal}^N denote the computed solution, i.e., $u_{\text{Gal}}^N \in S^N$ is defined by

$$B(u_{\text{Gal}}^N, v^N) = (\varepsilon\nabla u_{\text{Gal}}^N, \nabla v^N) + (\mathbf{a}\cdot\nabla u_{\text{Gal}}^N, v^N) + (bu_{\text{Gal}}^N, v^N) = (f, v^N)$$

for all $v^N \in S^N$. Existence and uniqueness of u_{Gal}^N follows from (6.12).

Theorem 6.36. *There exists a constant C such that*

$$\|u - u_{\text{Gal}}^N\|_{1,\varepsilon} \le CN^{-1}\ln N.$$

Proof. It is easy to see that for all $v^N \in S^N$ one has $B(u, v^N) = (f, v^N)$, so the Galerkin orthogonality property $B(u - u_{\text{Gal}}^N, v^N) = 0$ holds true. Let $u^I \in S^N$ be the nodal interpolant of u. By the coercivity inequality (6.12) and Galerkin orthogonality, we have

$$\begin{aligned}\min\{1, C_5\}\|u^I - u_{\text{Gal}}^N\|_{1,\varepsilon}^2 &\le B(u^I - u_{\text{Gal}}^N, u^I - u_{\text{Gal}}^N)\\ &= B(u^I - u, u^I - u_{\text{Gal}}^N).\end{aligned} \tag{6.38}$$

Now

$$
\begin{aligned}
|B(u^I - u, u^I - u^N_{\text{Gal}})| &= \big|\big(- \varepsilon\nabla(u^I - u), \nabla(u^I - u^N_{\text{Gal}})\big) \\
&\quad + \big(\mathbf{a}\cdot\nabla(u^I - u) + b(u^I - u), u^I - u^N_{\text{Gal}}\big)\big| \\
&= \big|\big(- \varepsilon\nabla(u^I - u), \nabla(u^I - u^N_{\text{Gal}})\big) \\
&\quad - \big(u^I - u, \mathbf{a}\cdot\nabla(u^I - u^N_{\text{Gal}})\big) \\
&\quad + (b - \operatorname{div}\mathbf{a})(u^I - u, u^I - u^N_{\text{Gal}})\big| \\
&\le C\,\big[\|u - u^I\|_{1,\varepsilon}\cdot\|u^I - u^N_{\text{Gal}}\|_{1,\varepsilon} \\
&\quad + \|u - u^I\|_{L_2(\Omega_{11})}\|\mathbf{a}\cdot\nabla(u^I - u^N_{\text{Gal}})\|_{L_2(\Omega_{11})} \\
&\quad + \|u - u^I\|_{L_\infty(\Omega\setminus\Omega_{11})}\|\mathbf{a}\cdot\nabla(u^I - u^N_{\text{Gal}})\|_{L_1(\Omega\setminus\Omega_{11})}\big]\,. \qquad (6.39)
\end{aligned}
$$

A standard inverse inequality on the uniform coarse mesh on Ω_{11} yields

$$\|\mathbf{a}\cdot\nabla(u^I - u^N_{\text{Gal}})\|_{L_2(\Omega_{11})} \le CN\|u^I - u^N_{\text{Gal}}\|_{L_2(\Omega_{11})}. \qquad (6.40)$$

The choice of λ_x and λ_y in the Shishkin mesh implies that

$$\text{measure of } (\Omega\setminus\Omega_{11}) \le C\varepsilon\ln N,$$

so a Cauchy–Schwarz inequality gives

$$
\begin{aligned}
\|\mathbf{a}\cdot\nabla(u^I - u^N_{\text{Gal}})\|_{L_1(\Omega\setminus\Omega_{11})} &\le (C\varepsilon\ln N)^{1/2}\|\mathbf{a}\cdot\nabla(u^I - u^N_{\text{Gal}})\|_{L_2(\Omega\setminus\Omega_{11})} \\
&\le C(\ln N)^{1/2}\|u^I - u^N_{\text{Gal}}\|_{1,\varepsilon}. \qquad (6.41)
\end{aligned}
$$

Substituting (6.39)–(6.41) into (6.38) and cancelling $\|u^I - u^N_{\text{Gal}}\|_{1,\varepsilon}$, we obtain

$$
\begin{aligned}
\|u^I - u^N_{\text{Gal}}\|_{1,\varepsilon} &\le C\Big[\|u - u^I\|_{1,\varepsilon} + N\|u - u^I\|_{L_2(\Omega_{11})} \\
&\qquad + (\ln N)^{1/2}\|u - u^I\|_{L_\infty(\Omega\setminus\Omega_{11})}\Big] \\
&\le CN^{-1}\ln N
\end{aligned}
$$

by (6.37). Now invoke (6.37b) and a triangle inequality to complete the proof. □

The bound of Theorem 6.36 is sharpened to $\mathcal{O}\left(N^{-2}\ln^2 N\right)$ in [**Lin00**, **Zha03**], but there are some differences in the FEM and norms in these two papers.

Despite these theoretical convergence results, the standard Galerkin FEM on a Shishkin mesh is not a good numerical method in practice because the discrete matrix is ill-conditioned and expensive to invert; see [**LS01b**].

Exercise 6.37. In the derivation of (6.39) in the proof of Theorem 6.36, what goes wrong if one applies a Cauchy–Schwarz inequality to estimate $\big(u^I - u, \mathbf{a}\cdot\nabla(u^I - u^N_{\text{Gal}})\big)$ on $\Omega\setminus\Omega_{11}$, as we did on Ω_{11}?

Remark 6.38. For the reaction-diffusion problem discussed in Remark 6.7, Liu et al. [**LMSZ09**, Theorem 3.1] derive a result analogous to Theorem 6.36 using the bounds described in Remark 4.18, and in [**RS15b**] a convergence result on a Shishkin mesh is proved in a balanced norm.

6.6.2. SDFEM with bilinears. We consider the same problem as in section 6.6.1, i.e., a problem posed on the unit square whose solution has two exponential outflow boundary layers and a corner layer, but no characteristic layers. For bilinears on a rectangular Shishkin mesh, it is shown in [**ST03**] that $\|u - u_{SD}\|_0 \leq C[\varepsilon N^{-3/2} + N^{-2}(\ln N)^2]$, which under the reasonable additional assumption that $\varepsilon \leq N^{-1/2}$ yields $\|u - u_{SD}\|_0 \leq CN^{-2}(\ln N)^2$ (optimal up to the $\ln N$ factor). This result relies on certain interpolation error identities [**Lin91**] that enable the analysis to be carried out separately on each mesh rectangle and yield a higher order of convergence than a direct application of bounds like (6.36).

To communicate the complexity of the analysis, we shall list only the main steps without detailed proofs, which can be found in [**ST03**].

Recall the decomposition $u = S + E_{21} + E_{12} + E_{22}$ stated in (4.14).

Using (6.36) and some other ideas, one can show the following result.

Lemma 6.39. *Let S^I and E^I denote the piecewise bilinear interpolants of S and E, respectively, on the Shishkin mesh Ω_S^N, where the function E can be any one of E_1, E_2, or E_{12}. Then there exists a constant C such that the following interpolation error estimates hold true:*

$$\|S - S^I\|_{0,\Omega} \leq C\,N^{-2}, \tag{6.42a}$$

$$\|E\|_{0,\Omega_{11}} \leq C\,\varepsilon^{1/2} N^{-5/2}, \tag{6.42b}$$

$$\varepsilon\|\Delta E\|_{L^1(\Omega_{11})} + \|\nabla E\|_{L^1(\Omega_{11})} \leq C\,N^{-5/2}, \tag{6.42c}$$

$$N^{-1}\|\nabla E^I\|_{0,\Omega_{11}} + \|E^I\|_{0,\Omega_{11}} \leq C\left(\varepsilon^{1/2}N^{-5/2} + N^{-3}\right), \tag{6.42d}$$

$$\|E - E^I\|_{0,\Omega} \leq C(N^{-}\ln N)^2. \tag{6.42e}$$

Exercise 6.40. Prove the inequalities (6.42) for the case $E = E_1$. *Hints.* Inequality (6.42a) is a standard classical result. The estimates (6.42b) and (6.42c) are straightforward to derive using the bounds on E_1 from (4.14). Inequality (6.42d) is more difficult to prove. To obtain (6.42e), use (6.42b) and (6.42d) for $\|E - E^I\|_{0,\Omega_{11}}$ and some calculations invoking (6.36) for $\|E - E^I\|_{0,\Omega\setminus\Omega_{11}}$.

Write the dimensions of each mesh rectangle τ as $h_{x,\tau}$ by $h_{y,\tau}$, and denote its barycentre by (x_τ, y_τ). Set

$$G_\tau(x) = \frac{1}{2}\left[(x - x_\tau)^2 - \left(\frac{h_{x,\tau}}{2}\right)^2\right], \quad F_\tau(y) = \frac{1}{2}\left[(y - y_\tau)^2 - \left(\frac{h_{y,\tau}}{2}\right)^2\right].$$

Denote the east, north, west, and south edges of τ by $l_{i,\tau}$ for $i = 1, \ldots, 4$, respectively.

Next, we state the Lin identities for bilinear interpolants. Each one can be proved by starting from the right-hand side of the identity: Since $(w^I)_{xx}$, $(w^I)_{yy}$ and all third-order derivatives of w^I vanish, these terms can be introduced at appropriate places in the right-hand side. Then one integrates by parts and takes into consideration the definitions of F_τ and G_τ. These identities appeared first in [**Lin91**]; for details of their proof a more convenient source is [**Zha03**].

Lemma 6.41. *Let τ be a mesh rectangle. Let $w \in H^3(\tau)$, and let $w^I \in Q_1(\tau)$ be its bilinear interpolant. Then for each $v^N \in Q_1(\tau)$ one has*

$$\int_\tau (w - w^I)_x v_x^N \,dx\,dy = \int_\tau w_{xyy}\left(F_\tau v_x^N - \frac{1}{3}\left(F_\tau^2\right)' v_{xy}^N\right) dxdy,$$

$$\begin{aligned}\int_\tau (w - w^I)_x v_y^N \,dx\,dy = &\int_\tau \left(F_\tau w_{xyy}(v_y^N - G_\tau' v_{xy}^N) + G_\tau w_{xxy} v_x^N\right) dx\,dy \\ &- \int_{l_{2,\tau}} G_\tau w_{xx} v_x^N \,dx + \int_{l_{4,\tau}} G_\tau w_{xx} v_x^N \,dx,\end{aligned}$$

$$\begin{aligned}\int_\tau (w - w^I)_y v_x^N \,dx\,dy = &\int_\tau \left(G_\tau w_{xxy}(v_x^N - F_\tau' v_{xy}^N) + F_\tau w_{xyy} v_y^N\right) dx\,dy \\ &- \int_{l_{1,\tau}} F_\tau w_{yy} v_y^N \,dy + \int_{l_{3,\tau}} F_\tau w_{yy} v_y^N \,dy,\end{aligned}$$

$$\int_\tau (w - w^I)_y v_y^N \,dxdy = \int_\tau w_{xxy}\left(G_\tau v_y^N - \frac{1}{3}\left(G_\tau^2\right)' v_{xy}^N\right) dx\,dy.$$

Remark 6.42. For a way of avoiding the derivation of ingenious identities like those of Lemma 6.41 yet still obtaining optimal-order error bounds, see [**DLP12**].

With the help of Lemma 6.41, some detailed calculations yield the next lemma (here we use the standard Sobolev space notation $W^{m,p}$).

Lemma 6.43. *Let $\varphi \in W^{1,\infty}(\Omega)$ satisfy $\|\varphi\|_{W^{1,\infty}} \leq C$ for some constant C. Let $w^I \in S^N$ be the piecewise bilinear interpolant of $w \in H^3(\Omega) \cap W^{2,\infty}(\Omega)$*

on the Shishkin mesh Ω_S^N. Then for all $v^N \in Q_1(K)$ we have

$$\left|\int_{\Omega_{11}} \varphi(w-w^I)_x v_x^N \,dx\,dy\right| \le C\,N^{-2}\left(|w|_2+|w|_3\right)\|v_x^N\|_0,$$

$$\left|\int_{\Omega_{11}} \varphi(w-w^I)_x v_y^N \,dx\,dy\right| \le C\,N^{-2}\left(|w|_{W^{2,\infty}}+\|w\|_3\right)\left(\|v_y^N\|_0+\|v_x^N\|_0\right) + C\varepsilon^{1/2}N^{-2}(\ln N)^{1/2}|w|_{W^{2,\infty}}\|v_{xy}^N\|_0,$$

$$\left|\int_{\Omega_{11}} \varphi(w-w^I)_y v_x^N \,dx\,dy\right| \le C\,N^{-2}\left(|w|_{W^{2,\infty}}+\|w\|_3\right)\left(\|v_x^N\|_0+\|v_y^N\|_0\right) + C\varepsilon^{1/2}N^{-2}(\ln N)^{1/2}|w|_{W^{2,\infty}}\|v_{xy}^N\|_0,$$

$$\left|\int_{\Omega_{11}} \varphi(w-w^I)_y v_y^N \,dx\,dy\right| \le C\,N^{-2}\left(|w|_2+|w|_3\right)\|v_y^N\|_0.$$

Exercise 6.44. For simplicity take $\varphi \equiv 1$. Use Lemma 6.41 to prove that

$$\left|\int_{\Omega_{11}} (w-w^I)_x v_x^N \,dx\,dy\right| \le C\,N^{-2}\left(|w|_2+|w|_3\right)\|v_x^N\|_0.$$

Show that a direct application of the Cauchy–Schwarz inequality, without appealing to Lemma 6.41, yields only an $\mathcal{O}\left(N^{-1}\right)$ bound:

$$\left|\int_{\Omega_{11}} (w-w^I)_x v_x^N \,dx\,dy\right| \le C\,N^{-1}|w|_2\,\|v_x^N\|_0.$$

Choose the piecewise-constant streamline diffusion parameter (cf. (6.25)) as follows:

$$\delta_\tau = \begin{cases} N^{-1} & \text{if } \tau\subset\Omega_{11} \text{ and } \varepsilon\le N^{-1},\\ \varepsilon^{-1}N^{-2} & \text{if } \tau\subset\Omega_{11} \text{ and } \varepsilon> N^{-1},\\ 0 & \text{otherwise.}\end{cases}$$

Earlier results are used to derive the next two lemmas.

Lemma 6.45. *For all $v^N \in S^N$, we have*

$$\left|B(u-u^I,v^N)\right| \le C\left[\varepsilon N^{-3/2}+(N^{-1}\ln N)^2\right]\|v^N\|_{1,\varepsilon}.$$

Write $B_{SD}(\cdot,\cdot) = B(\cdot,\cdot)+B_{\text{stab}}(\cdot,\cdot)$ as in (6.17).

Lemma 6.46. *For some constant C one has*

$$\left|B_{\text{stab}}(u-u^I,v^N)\right| \le CN^{-2}(\ln N)^{1/2}\|v^N\|_{SD} \quad \forall v^N\in S^N.$$

Exercise 6.47. Use Lemma 6.39 and an inverse inequality to prove

$$\left|B_{\text{stab}}(E_1-E_1^I,v^N)\right| \le CN^{-2}\|v^N\|_{SD} \quad \forall v^N\in S^N.$$

It is now straightforward to prove our main result.

Theorem 6.48. *The piecewise bilinear SDFEM solution u_{SD}^N satisfies*

$$\|u_{SD}^N - u^I\|_{SD} \le C\left[\varepsilon N^{-3/2} + (N^{-1}\ln N)^2\right].$$

Proof. By Lemma 6.15 (recall Remark 6.16) and (6.18), we have

$$\begin{aligned}\frac{1}{2}\|u_{SD}^N - u^I\|_{SD}^2 &\le B_{SD}(u_{SD}^N - u^I, u_{SD}^N - u^I)\\ &= B_{SD}(u - u^I, u_{SD}^N - u^I)\\ &= B(u - u^I, u_{SD}^N - u^I) + B_{\text{stab}}(u - u^I, u_{SD}^N - u^I).\end{aligned}$$

Now invoke Lemmas 6.45 and 6.46 to complete the proof. □

Remark 6.49. While Theorem 6.48 was proved in 2003 in [**ST03**], only recently in [**ZLY16**] was an almost-optimal L_2 result similar to this theorem obtained for piecewise *linears* in the SDFEM applied to the same convection-diffusion problem on a *triangular* Shishkin mesh. The main difference in the analysis in this case is that for linears on triangles there are no interpolation error identities on an individual triangle that are analogous to those of Lemma 6.41 for bilinears on a rectangle. Instead, one must group triangles in pairs to obtain an analogue of Lemma 6.41. In [**ZLY16**] some new interpolation error results for linears are derived, and it is then shown that

$$\|u - u_{SD}\|_0 \le C[\varepsilon^{1/2}N^{-3/2} + N^{-2}(\ln N)^{5/2}],$$

which, under the reasonable additional assumption that $\varepsilon \le N^{-1}$, yields $\|u - u_{SD}\|_0 \le CN^{-2}(\ln N)^{5/2}$.

Remark 6.50. In [**LS01b**] numerical experiments compare several methods on the same Shishkin mesh for our usual test problem on the unit square whose solution has exponential outflow layers along $x = 1$ and $y = 1$ and can be decomposed as in (4.14). The methods considered are central differencing, simple upwinding, the hybrid difference scheme of [**LS99**], defect correction using simple upwinding and central differencing (see Remark 5.4), linear and bilinear Galerkin FEMs (see section 6.6.1 and [**Zha03**]), and the linear and bilinear SDFEM (see section 6.6.2). Graphs of the computed solutions, errors, and convergence rates in the discrete $L_\infty(\Omega)$ norm are given, and known theoretical convergence results for the various methods are listed. It is concluded that, taking into account any difficulties that arise in solving the discrete linear systems, the methods that performed best for this problem are the defect correction method and the two SDFEMs, and that inside the layers bilinears are more accurate than linears.

In a similar spirit, the more recent paper [**ACF+11**] compares numerically several stabilized FEMs that are used to solve the Hemker problem described in section 4.2.2. The conclusions/recommendations of this useful paper are lengthy, and we do not reproduce them here.

Remark 6.51 (Supercloseness and postprocessing)**.** Theorem 6.48 implies that the piecewise bilinear solution u_{SD}^N on a Shishkin mesh satisfies

$$\|u_{SD}^N - u^I\|_{1,\varepsilon} \le \|u_{SD}^N - u^I\|_{SD} \le C\left[\varepsilon N^{-3/2} + (N^{-1}\ln N)^2\right],$$

while (6.37b) then implies that

$$\|u_{SD}^N - u\|_{1,\varepsilon} \le \|u_{SD}^N - u^I\|_{1,\varepsilon} + \|u^I - u\|_{1,\varepsilon} \le CN^{-1}\ln N.$$

This phenomenon, where $\|u_{SD}^N - u^I\| \ll \|u_{SD}^N - u\|$ in some norm $\|\cdot\|$, is called *supercloseness* of u_{SD}^N. On a Shishkin mesh it can be exploited easily and cheaply to get a more accurate approximation of u by *postprocessing* u_{SD}^N: form a macromesh by taking groups of four neighbouring mesh rectangles which each have nine nodes of the original mesh, construct the piecewise biquadratic interpolant Pu_{SD}^N of u_{SD}^N at these nodes, and then

$$\|Pu_{SD}^N - u\|_{1,\varepsilon} \le C\left[\varepsilon N^{-3/2} + (N^{-1}\ln N)^2\right];$$

see [**ST03**] for more details.

6.7. Discontinuous Galerkin finite element method

In recent years, several versions of the *discontinuous Galerkin FEM* (dGFEM) have attracted a great deal of attention in the research literature. Like the SDFEM, the dGFEM achieves stability by a judicious choice of bilinear form, but the details of the construction are very different from section 6.4.

The method's name comes from its use of a standard piecewise polynomial trial space that is not required to be continuous across element boundaries. This local nature means the method is more readily parallelizable than (say) the SDFEM, and it clearly permits the use of polynomials of different degrees on different elements, which can be exploited to gain increased accuracy when the problem is quite smooth on only part of the domain—as is usually the case with convection-diffusion problems. A drawback is the much larger number of degrees of freedom compared with finite element spaces that lie in $C(\Omega)$.

Methods of this type were first introduced in the 1970s, and today there are several prominent variants. Arnold et al. [**ABCM02**] consider the problem $-\Delta u = f$ on Ω with $u = 0$ on $\partial\Omega$ and show that nine distinct versions of the dGFEM can be placed in the framework of a mixed-method weak formulation. They go on to analyse the stability of these methods, but this is of limited value in the context of convection-diffusion problems where the Laplacian is multiplied by a small parameter. This paper also gives an account of the historical development of dGFEMs that includes methods specifically designed for convection-diffusion problems.

To introduce the stabilization idea used in most dGFEMs, we describe an older and simpler numerical technique known as *Nitsche's method*, which imposes boundary conditions only weakly.

Let $\Omega \subset \mathbb{R}^2$ be a bounded domain with $\mathbf{n}$ the outward-pointing unit normal to $\partial\Omega$.

Example 6.52 (Nitsche's method). Consider a Dirichlet boundary value problem based on Poisson's equation,

$$-\Delta u = f \text{ in } \Omega, \quad u = g \text{ on } \partial\Omega. \tag{6.43}$$

As usual in FEMs, we convert the differential equation to a weak form by multiplying it by an arbitrary test function $v \in H^1(\Omega)$ and then integrating by parts:

$$\begin{aligned}(f, v) = (-\Delta u, v) &= (\nabla u, \nabla v) - \int_{\partial\Omega} (\nabla u \cdot \mathbf{n})v \\ &= (\nabla u, \nabla v) - \int_{\partial\Omega} (\nabla u \cdot \mathbf{n})v - \int_{\partial\Omega} (\nabla v \cdot \mathbf{n})(u - g), \end{aligned} \tag{6.44}$$

where the last integral adds zero to the calculation because $u = g$ on $\partial\Omega$. It was introduced because (6.44) can now be written as

$$B_1(u, v) := (\nabla u, \nabla v) - \int_{\partial\Omega} (\nabla u \cdot \mathbf{n})v - \int_{\partial\Omega} (\nabla v \cdot \mathbf{n})u = (f, v) - \int_{\partial\Omega} (\nabla v \cdot \mathbf{n})g$$

for all $v \in H^1(\Omega)$, where $B_1(\cdot, \cdot)$ is a symmetric bilinear form.

But we cannot prove that $B_1(\cdot, \cdot)$ is coercive over $H^1(\Omega) \times H^1(\Omega)$. To obtain this very desirable property, we add another term to the bilinear form—while maintaining its symmetry. For arbitrary $\mu > 0$, add the term

$$\mu \int_{\partial\Omega} (u - g)v$$

to (6.44). This equation can then be rearranged as

$$\begin{aligned}B_2(u, v) &:= (\nabla u, \nabla v) - \int_{\partial\Omega} (\nabla u \cdot \mathbf{n})v - \int_{\partial\Omega} (\nabla v \cdot \mathbf{n})u + \mu \int_{\partial\Omega} uv \\ &= (f, v) - \int_{\partial\Omega} (\nabla v \cdot \mathbf{n})g + \mu \int_{\partial\Omega} gv\end{aligned}$$

for all $v \in H^1(\Omega)$. Replacing u, v by u^h, v^h from a standard piecewise polynomial finite element space $S^h \subset H^1(\Omega)$ on a quasi-uniform mesh of width h, one can show (cf. [**Sch08**]) that if $\mu = Ch^{-1}$ for sufficiently large C, then $B_2(\cdot, \cdot)$ is coercive over $S^h \times S^h$ with respect to a certain mesh-dependent norm.

Then B_2 is symmetric, coercive and by its construction enjoys the Galerkin orthogonality property $B(u - u^h, v^h) = 0$ for all $v^h \in S^h$.

The message of Example 6.52 is that one can choose to impose boundary conditions weakly and still obtain a symmetric and coercive bilinear form, provided one is willing to include a user-chosen *penalty parameter* (viz., μ above) in the weak formulation of the problem.

Exercise 6.53. Prove the assertion of Example 6.52 that if $\mu = Ch^{-1}$ for sufficiently large C, then $B_2(\cdot, \cdot)$ is coercive over $S^h \times S^h$ with respect to a certain mesh-dependent norm. You will need a standard trace inequality that is not stated in our book.

Remark 6.54. In [**Sch08**] a related method for convection-diffusion problems is investigated. A nonsymmetric version of Nitsche's method without any penalty parameter is analysed in [**Bur12**].

Given the diversity of methods described as dGFEMs, we shall not attempt to give a thorough survey of this area. Instead we concentrate on one variant, and the references appearing in this section will assist the reader who wishes to broaden her or his knowledge of the dGFEM.

Thus we consider in detail the version of the nonsymmetric interior penalty dGFEM (NIPD) from [**HSS02**]; related methods appear in, e.g., [**OBB98**] and [**RWG01**].

Assume that Ω is polygonal. Let $\mathcal{T}$ be a partition of Ω into elements κ (e.g., triangles or rectangles). In [**HSS02**] up to one hanging node is permitted for each κ, but for simplicity we shall assume that our partition has no hanging nodes. Assume also that each $\kappa \in \mathcal{T}$ is an affine image of a fixed master element $\hat{\kappa}$, i.e., that $\kappa = F_\kappa(\hat{\kappa})$ where $\hat{\kappa}$ is either the open unit simplex or the open unit square in $\mathbb{R}^2$. For each nonnegative integer k, let $\mathcal{P}_k(\hat{\kappa})$ denote the set of polynomials of total degree k on $\hat{\kappa}$. (If $\hat{\kappa}$ is the unit square, one can also consider $\mathcal{Q}_k(\hat{\kappa})$, the set of all tensor-product polynomials on $\hat{\kappa}$ of degree k in each coordinate direction.) For each $\kappa \in \mathcal{T}$ write p_κ for the local polynomial degree. Set $\mathbf{p} = \{p_\kappa : \kappa \in \mathcal{T}\}$ and $\mathbf{F} = \{F_\kappa : \kappa \in \mathcal{T}\}$ and define the finite element space

$$S^{\mathbf{p}}(\Omega, \mathcal{T}, \mathbf{F}) = \{v \in L_2(\Omega) : v|_\kappa \circ F_\kappa \in \mathcal{R}_{p_\kappa}(\hat{\kappa})\},$$

where $\mathcal{R}$ is either $\mathcal{P}$ or $\mathcal{Q}$.

For $s = 0, 1$ define the broken Sobolev spaces

$$H^s(\Omega, \mathcal{T}) = \{v \in L_2(\Omega) : v|_\kappa \in H^s(\kappa) \ \forall \kappa \in \mathcal{T}\}.$$

Let $\partial\kappa$ denote the boundary of κ for each $\kappa \in \mathcal{T}$. Define the inflow and outflow parts of $\partial\kappa$ by

$$\partial^-\kappa = \{(x,y) \in \partial\kappa : \mathbf{a}(x,y)\cdot \mathbf{n}_\kappa(x,y) < 0\},$$
$$\partial^+\kappa = \{(x,y) \in \partial\kappa : \mathbf{a}(x,y)\cdot \mathbf{n}_\kappa(x,y) \geq 0\},$$

respectively, where $\mathbf{n}_\kappa(x,y)$ denotes the outward-pointing unit normal to $\partial\kappa$ at $(x,y)\in\partial\kappa$.

Let $v \in H^1(\Omega,\mathcal{T})$. For each $\kappa \in \mathcal{T}$, denote by v_κ^+ the inner trace of $v|_\kappa$ on $\partial\kappa$. If $\partial^-\kappa\backslash\partial\Omega$ is nonempty, then for almost every point $(x,y)\in\partial^-\kappa\backslash\partial\Omega$ there exists a unique $\kappa' \in \mathcal{T}$ (which depends on (x,y)) such that $x \in \partial^+\kappa'$ and $\kappa' \cap (\partial^-\kappa \setminus \partial\Omega)$ has nonzero one-dimensional measure, and we define the outer trace v_κ^- of v on $\partial^-\kappa\setminus\partial\Omega$ relative to κ to be the inner trace $v_{\kappa'}^+$ relative to κ'. Then define the jump of v across $\partial^-\kappa\backslash\partial\Omega$ by $\lfloor v \rfloor_\kappa = v_\kappa^+ - v_\kappa^-$.

We shall drop the subscript κ from the above notation when it is clear from the context what is intended.

Let $\mathcal{E}_{\text{int}}$ be the set of all open one-dimensional edges of the partition $\mathcal{T}$ that lie in Ω. Set $\Gamma_{\text{int}} = \{x \in \Omega : x \in e \text{ for some } e \in \mathcal{E}_{\text{int}}\}$. Numbering the elements κ consecutively, for each $e \in \mathcal{E}_{\text{int}}$ there exist indices i and j such that $i > j$ and the elements κ_i and κ_j share the interface e. Define the (element-numbering-dependent) jump of $v \in H^1(\Omega,\mathcal{T})$ across e and the mean value of v on e by

$$[v]_e = v|_{\partial\kappa_i \cap e} - v|_{\partial\kappa_j\cap e} \quad \text{and} \quad \langle v\rangle_e = \frac{1}{2}\left(v|_{\partial\kappa_i\cap e} + v|_{\partial\kappa_j\cap e}\right),$$

respectively. Furthermore, for each $e \in \mathcal{E}_{\text{int}}$ let $\mathbf{n}_{ij}$ denote the unit vector that is normal to e and pointing from κ_i to κ_j; if $e \subset \partial\Omega$, take $\mathbf{n}_{ij} = \mathbf{n}$.

The bilinear form associated with the NIPD for (4.1a) with $u \equiv 0$ on $\partial\Omega$ is

$$\begin{aligned} B_{DG}(v,w) = \sum_{\kappa\in\mathcal{T}} \Bigg(\varepsilon \int_\kappa \nabla v\cdot\nabla w + \int_\kappa (\mathbf{a}\cdot\nabla v + bv)w \\ - \int_{\partial^-\kappa\cap\partial^-\Omega} (\mathbf{a}\cdot\mathbf{n}_\kappa) v^+ w^+ - \int_{\partial^-\kappa\backslash\partial\Omega} (\mathbf{a}\cdot\mathbf{n}_\kappa)\lfloor v\rfloor w^+ \Bigg) \\ + \varepsilon \int_{\partial\Omega} (v(\nabla w\cdot\mathbf{n}) - (\nabla v\cdot\mathbf{n})w) + \int_{\partial\Omega} \sigma v w \\ + \varepsilon \int_{\Gamma_{int}} ([v]\langle\nabla w\cdot\mathbf{n}\rangle - \langle\nabla v\cdot\mathbf{n}\rangle[w]) + \int_{\Gamma_{int}} \sigma[v][w], \end{aligned}$$

for all $v, w \in H^1(\Omega,\mathcal{T})$. Here σ, the user-chosen nonnegative *discontinuity-penalization parameter*, is defined by

$$\sigma|_e = \sigma_e \quad \text{for each } e \in \mathcal{E}_{\text{int}} \cup \partial\Omega.$$

When solving (4.1), Houston et al. [**HSS02**] choose $\sigma_e = \mathcal{O}(\varepsilon/h_e)$ where h_e is the length of edge e.

The NIPD method is then as follows: Find $u_{dG} \in S^{\mathbf{p}}(\Omega, \mathcal{T}, \mathbf{F})$ such that

$$B(u_{dG}, w^N) = \sum_{\kappa \in \mathcal{T}} \int_\kappa f w^N \quad \text{for all } w^N \in S^{\mathbf{p}}(\Omega, \mathcal{T}, \mathbf{F}). \tag{6.45}$$

Existence and uniqueness of a solution to (6.45) are shown in [**HSS02**].

Assuming that $u \in H^2(\Omega, \mathcal{T})$ and ∇u is continuous across each edge $e \in \mathcal{E}_{\text{int}}$, one can deduce the Galerkin orthogonality property

$$B_{dG}(u - u_{dG}, w^N) = 0 \quad \forall w^N \in S^{\mathbf{p}}(\Omega, \mathcal{T}, \mathbf{F}).$$

For all $v \in H^2(\Omega, \mathcal{T})$ define the norm $\|\cdot\|_{dG}$ by $\|v\|_{dG}^2 = B_{dG}(v, v)$. Setting $(v, w)_e = \int_e |\mathbf{a} \cdot \mathbf{n}_\kappa| vw$ for each $e \subset \partial\kappa$ and $\|v\|_e^2 = (v, v)_e$, after some manipulation one gets

$$\begin{aligned} \|v\|_{dG}^2 = &\sum_{\kappa \in \mathcal{T}} \left(\varepsilon \|\nabla v\|_{0,\kappa}^2 + \|c_0 v\|_{0,\kappa}^2\right) + \int_{\partial\Omega} \sigma v^2 + \int_{\Gamma_{int}} \sigma [v]^2 \\ &+ \frac{1}{2} \sum_{\kappa \in \mathcal{T}} \left(\|v^+\|_{\partial_- \kappa \cap \partial\Omega}^2 + \|v^+ - v^-\|_{\partial_- \kappa \setminus \partial\Omega}^2 + \|v^+\|_{\partial_+ \kappa \cap \partial\Omega}^2 \right), \end{aligned}$$

where $\|\cdot\|_{0,\kappa}$ is the $L^2(\kappa)$ norm and we set $c_0(x, y) = \sqrt{b(x, y) - \operatorname{div} \mathbf{a}(x, y)/2}$; by (6.9) the function c_0 is well-defined. Clearly, $\|\cdot\|_{dG}$ is stronger than $\|\cdot\|_{1,\varepsilon}$.

Now [**HSS02**] write $u - u_{dG} = (u - \Pi u) + (\Pi u - u_{dG})$ where Π is the orthogonal projector in L_2 into $S^{\mathbf{p}}(\Omega, \mathcal{T}, \mathbf{F})$. From Galerkin orthogonality one has

$$\|\Pi u - u_{dG}\|_{dG}^2 = B_{dG}(\Pi u - u_{dG}, \Pi u - u_{dG}) = B_{dG}(\Pi u - u, \Pi u - u_{dG}), \tag{6.46}$$

and, under the assumption that $\mathbf{a} \cdot \nabla w^N|_\kappa$ lies in $S^{\mathbf{p}}(\Omega, \mathcal{T}, \mathbf{F})$ for all $w^N \in S^{\mathbf{p}}(\Omega, \mathcal{T}, \mathbf{F})$, some analysis of the right-hand side of (6.46) enables $\|\Pi u - u_{dG}\|_{dG}$ to be estimated in terms of various norms of $u - \Pi u$. Invoking the triangle inequality $\|u - u_{dG}\|_{dG} \le \|u - \Pi u\|_{dG} + \|\Pi u - u_{dG}\|_{dG}$ then leads to a bound on $\|u - u_{dG}\|_{dG}$. In the particular case where the mesh elements are rectangles, piecewise polynomials of degree k are used, the mesh diameter is h, and the solution u lies in $H^{k+1}(\Omega)$, the bound becomes

$$\|u - u_{dG}\|_{dG} \le C(\varepsilon^{1/2} h^k + h^{k+1/2}) \|u\|_{H^{k+1}(\Omega)}, \quad \text{where } C = C(k).$$

Note that the right-hand side here depends on a Sobolev norm of u that is typically $\mathcal{O}\left(\varepsilon^{-k-1/2}\right)$. It may be possible to use cut-off functions to localize this result away from layers, removing this undesirable feature.

The above analysis from [**HSS02**] assumes that the mesh is shape-regular, so it excludes the long thin elements present in any good mesh that is specifically designed to improve the behaviour of the method inside layers. Roos and Zarin [**RZ03**] applied this dGFEM to a problem on the unit square that has exponential layers along $x = 1$ and $y = 1$ and no other layers. Working with piecewise bilinears on a rectangular Shishkin mesh like that of Figure 5.2 with N mesh intervals in each coordinate direction, they adapt the analysis of [**HSS02**] to this situation (which entails a different choice for σ_e on part of the mesh) and prove that

$$\|u - u_{dG}\|_{dG} \leq CN^{-1}(\ln N)^{3/2}. \tag{6.47}$$

A related paper [**ZR05**] considers a problem similar to Example 4.2 and, using a Shishkin mesh similar to the one in Figure 5.3 with N mesh intervals in each coordinate direction, again obtains the bound (6.47).

We remind the reader that there is no universal agreement on a "best" form of the dGFEM. For example, in [**GK03**] a symmetric version of our bilinear form $B_{dG}(v, w)$ is considered; it is obtained by changing the signs of the terms $\varepsilon \int_{\partial\Omega} v(\nabla w \cdot \mathbf{n}_{ij})$ and $\varepsilon \int_{\Gamma_{\text{int}}} [v]\langle \nabla w \cdot \mathbf{n}_{ij}\rangle$.

For further work in this area see [**DPE12, ZXZ09, Roo12**] and their references, and the discussion in [**RST08**, Section III. 3.4.3].

6.8. Continuous interior penalty (CIP) method

The *continuous interior penalty* (*CIP*) *method* is a general technique for the stabilization of finite element methods, irrespective of the cause of the instability. It dates from 1976 [**DD76**], and it was applied to convection-dominated problems by Burman and others in several papers, including [**BE06, BGL09, BH04, BS16, Bur05, Bur12**]. The analyses in these papers are for shape-regular meshes. Here we shall discuss the analysis of a CIP method on a Shishkin mesh (which of course is not shape-regular) for the problem of section 6.6, which is posed on the unit square and whose solution exhibits exponential boundary layers along the sides $x = 1$ and $y = 1$ of the domain. Our presentation is based on [**LSZ**].

Thus, consider the problem

$$-\varepsilon\Delta u + \mathbf{a}(x, y) \cdot \nabla u + b(x, y)u = f(x, y) \quad \text{on } \Omega = (0, 1)^2, \tag{6.48a}$$

$$u = 0 \quad \text{on } \partial\Omega, \tag{6.48b}$$

with $\mathbf{a} = (a_1, a_2) \geq (\alpha_1, \alpha_2) > (0, 0)$. Assume that the decomposition (4.14) of the solution u is valid.

Without loss of generality (recall Remark 4.15) one can also assume that

$$b(x, y) - \frac{\operatorname{div} \mathbf{a}(x, y)}{2} \geq C_5 > 0 \quad \text{on } \bar{\Omega} \text{ for some constant } C_5. \tag{6.49}$$

We shall use the rectangular Shishkin mesh of section 6.6 (see Figure 6.4 on p. 118) with the mesh transition points

$$\lambda_x = \frac{m\varepsilon}{\alpha_1} \ln N, \quad \lambda_y = \frac{m\varepsilon}{\alpha_2} \ln N,$$

where the user-chosen constant m satisfies $m \geq 5/2$ and there are N mesh intervals in each cooordinate direction. Our finite element space $S^N \subset H_0^1(\Omega)$ comprises globally continuous piecewise bilinears on this mesh. Our discussion will employ the regions $\Omega_{11}, \Omega_{12}, \Omega_{21}$, and Ω_{22} of Figure 6.4.

Recall the standard Galerkin bilinear form $B(\cdot,\cdot)$ of section 6.6.1. The CIP method modifies this bilinear form by adding terms that penalize jumps in derivatives of the computed solution across element edges; in this way it inhibits oscillations. It is defined as follows: Find $u^N \in S^N$ such that

$$a(u^N, v^N) = (f, v^N) \quad \forall v^N \in S^N, \tag{6.50}$$

where

$$a(u^N, v^N) := B(u^N, v^N) + J(u^N, v^N)$$

with

$$J(u^N, v^N) := \sum_{e \subset \Omega_{11}^\circ} \gamma N^{-2} \int_e [\nabla u^N] \cdot [\nabla v^N] \, ds,$$

where e is an edge of a mesh rectangle, Ω_{11}° denotes the interior of Ω_{11}, the positive penalty parameter γ is chosen by the user (the numerical experiments of [**LSZ**] use $\gamma = 1.0$), and the jump $[q]$ of a piecewise continuous function q over an edge e is defined, for each $x \in e$, by

$$[q](x) = \begin{cases} \lim_{t \to 0+} [q(x + tn_e) - q(x - tn_e)] & \text{if } e \not\subset \partial\Omega, \\ 0 & \text{if } e \subset \partial\Omega, \end{cases}$$

where we associate a unique unit normal vector n_e with e. In (6.50) no stabilization is applied where the mesh is fine because this seems to give the best numerical results; see [**Fra08**], where several variants of the CIP method are considered.

The stabilizing mechanism of the term $J(\cdot,\cdot)$ in (6.50) differs from upwind stabilizations such as SDFEM.

The natural norm associated with $a(\cdot,\cdot)$ is

$$\|w\|_{CIP} := \left\{\varepsilon |w|_1^2 + C_5 \|w\|^2 + J(w,w)\right\}^{1/2}, \tag{6.51}$$

which is well-defined for those $w \in H^1(\Omega)$ for which $J(w,w)$ is defined. Using (6.49), it is easy to derive the coercivity inequality

$$a(v^N, v^N) \geq \|v^N\|_{CIP}^2 \quad \forall v^N \in S^N. \tag{6.52}$$

It follows that u^N is well-defined by (6.50).

The regularity of the true solution u that is stated in (4.14) shows that ∇u has no jumps in Ω, so $J(u, v^N) = 0$ for all $v^N \in S^N$. Thus (6.48) and (6.50) give the Galerkin orthogonality property

$$a(u - u^N, v^N) = 0 \quad \forall v^N \in S^N. \tag{6.53}$$

The error analysis of stabilized FEMs for convection-dominated problems often seems to work best when carried out in the framework of the streamline diffusion norm $\|\cdot\|_{SD}$, even if at first sight the FEM has no apparent connection with the SDFEM. For example, this is true of local projection stabilization; see [**KT11**]. Here also we shall make use of the SD norm even though the CIP method seems very different from the SD-FEM.

Remark 6.55 (The relationship between the CIP and SD norms)**.** An investigation of the exact relationship between $\|\cdot\|_{CIP}$ and $\|\cdot\|_{SD}$ for continuous piecewise linears on uniform meshes of width h in one dimension is carried out in [**LSZ**, Section 4]. The norms are defined as

$$\|w\|_{SD} := \left\{\varepsilon|w|_1^2 + \|w\|_0^2 + h|w|_1^2\right\}^{1/2}$$

and

$$\|w\|_{CIP} := \left\{\varepsilon|w|_1^2 + \|w\|_0^2 + h^2\sum_{i=1}^{M-1}([w']|_{x_i})^2\right\}^{1/2},$$

where the uniform mesh is $x_0 < x_1 < \cdots < x_M$, with $h = 1/M$. Let W_0^M be the space of globally continuous piecewise linear functions defined on this mesh that vanish at x_0 and x_M. It is shown in [**LSZ**] that if $\varepsilon \le h$, then

$$\|w^M\|_{CIP} \le 2\|w^M\|_{SD} \le CM^{1/4}\|w^M\|_{CIP} \quad \forall w^M \in W_0^M, \tag{6.54}$$

where C is a constant independent of M and w^M; moreover, a carefully constructed example demonstrates that the factor $M^{1/4}$ here is needed. Thus in one dimension the CIP norm is in general strictly weaker than the SD norm, and one can expect the same to be true in two dimensions.

Exercise 6.56. Prove the first inequality in (6.54). (It follows easily from the definitions of the norms; one does not need to use $\varepsilon \le h$.)

Exercise 6.57. The aim of this exercise is to prove the second inequality in (6.54). First show by a direct calculation that

$$\frac{h}{6}\sum_{i=1}^{M-1} w^2(x_i) \le \|w\|_0^2 \le h\sum_{i=1}^{M-1} w^2(x_i) \quad \forall w \in W_0^M.$$

Use this inequality, the fact that w' is constant on each mesh interval, and $st \le (s^2 + t^2)/2$ to prove that

$$h|w|_1^2 \le 6h^{-1/2} \left[\|w\|_0^2 + \sum_{i=1}^{M-1} h^2 \left([w']|_{x_i} \right)^2 \right] \quad \forall w \in W_0^M.$$

Combine this inequality with $\varepsilon \le h$ to obtain the second inequality in (6.54).

Exercise 6.58. This exercise will show, by constructing a specific example, that the factor $M^{1/4}$ in (6.54) cannot in general be eliminated. Define the $(M-1) \times (M-1)$ tridiagonal matrix

$$Q = \begin{pmatrix} 2 & -1 & & & \\ -1 & 2 & \ddots & & \\ & \ddots & \ddots & \ddots & \\ & & \ddots & 2 & -1 \\ & & & -1 & 2 \end{pmatrix}.$$

Let λ_k be any eigenvalue of Q. Let $\mathbf{v} = (\mathbf{v_1}, \mathbf{v_2}, \dots, \mathbf{v_{M-1}})^{\mathbf{T}}$ be a corresponding unit eigenvector of Q, so $Q\mathbf{v} = \lambda_{\mathbf{k}}\mathbf{v}$ and $\sum_{i=1}^{M-1} v_i^2 = 1$. Define $v \in W_0^M$ by $v(t_i) = v_i$ for $i = 1, 2 \dots, M-1$. By the first inequality in Exercise 6.57 we have $\|v\|_0^2 \le h$.

Next, summation by parts gives

$$h|v|_1^2 = \sum_{i=1}^{M} (v_i - v_{i-1})^2 = -\sum_{i=1}^{M-1} v_i(v_{i+1} - 2v_i + v_{i-1}) = \mathbf{v} \cdot Q\mathbf{v} = \lambda_k$$

and

$$h^2 \sum_{i=1}^{M-1} \left([v']|_{t_i} \right)^2 = \sum_{i=1}^{M-1} (v_{i+1} - 2v_i + v_{i-1})^2 = Q\mathbf{v} \cdot Q\mathbf{v} = \lambda_k^2.$$

Deduce that

$$\|v\|_{SD}^2 \ge \lambda_k \quad \text{and} \quad \|v\|_{CIP}^2 \le h + \frac{\varepsilon \lambda_k}{h} + \lambda_k^2.$$

We now choose a specific λ_k. Suppose that $M \ge 81$. Show that there exists an integer k^* satisfying $1 \le k^* \le M-1$ and $\sqrt{4/3}\, M^{3/4} \le k^* \le \sqrt{5/3}\, M^{3/4}$. The eigenvalues of Q are $\lambda_k = 2 - 2\cos(k\pi/M)$ for $k = 1, 2, \dots, M-1$. (*Exercise.* Find a source for this claim.) Show that

$$1 - \frac{t^2}{2!} \le \cos t \le 1 - \frac{t^2}{2!} + \frac{t^4}{4!} \quad \text{for } t \ge 0,$$

and deduce that

$$\frac{\pi^2}{M^{1/2}} \le \lambda_{k^*} \le \frac{5\pi^2}{3M^{1/2}},$$

and consequently (where $v^* \in W_0^M$ is associated with λ_{k^*})

$$\frac{\|v^*\|_{SD}^2}{\|v^*\|_{CIP}^2} \geq \frac{\pi^2}{M^{1/2}} \cdot \left[\frac{x_M - x_0}{M} + \frac{5\pi^2 \varepsilon M^{1/2}}{3(x_M - x_0)} + \frac{25\pi^4}{9M}\right]^{-1} \geq CM^{1/2}$$

if $\varepsilon \leq M^{-3/2}$, where C is some positive constant.

Returning to our two-dimensional problem, define the streamline diffusion norm by

$$\|v\|_{SD} := \left\{ \varepsilon |v|_1^2 + \|v\|_0^2 + \sum_{\tau \in \Omega_{11}} N^{-1} \|\mathbf{a} \cdot \nabla v\|_{0,\tau}^2 \right\}^{1/2},$$

where we have imitated the construction of the CIP method by introducing the streamline derivatives only on the coarse mesh Ω_{11}.

Lemma 6.59. *There exists a positive constant C such that*

$$\|v^N\|_{SD} \leq CN^{1/4} \|v^N\|_{CIP} \quad \forall v^N \in S^N.$$

In [**LSZ**] this two-dimensional analogue of the second inequality in (6.54) is deduced from (6.54).

Next we prove a technical result for the term $J(\cdot,\cdot)$ in the CIP norm. Set $e_{j;i} := \{x_i\} \times [y_j, y_{j+1}]$ and $e^{i;j} := [x_i, x_{i+1}] \times \{y_j\}$ for all i and j, and

$$\mathcal{E}_x = \{e_{j;i} : \ 0 \leq j < N/2, \ 1 \leq i < N/2\},$$
$$\mathcal{E}_y = \{e^{i;j} : \ 0 \leq i < N/2, \ 1 \leq j < N/2\}.$$

Thus $\mathcal{E}_x$ ($\mathcal{E}_y$) denotes the set of all interior edges of the mesh that lies in Ω_{11} and is perpendicular to the x-axis (to the y-axis). Each v_x^N (v_y^N) is continuous across each edge $e \in \mathcal{E}_y$ ($e \in \mathcal{E}_x$), so

$$([\nabla u^N] \cdot [\nabla v^N])\,|_e = \begin{cases} ([u_x^N] \cdot [v_x^N])\,|_e & \forall e \in \mathcal{E}_x, \\ ([u_y^N] \cdot [v_y^N])\,|_e & \forall e \in \mathcal{E}_y. \end{cases}$$

We therefore have

$$\begin{aligned} J(u^N, v^N) &= \left(\sum_{e \in \mathcal{E}_x} + \sum_{e \in \mathcal{E}_y} \right) \gamma N^{-2} \int_e [\nabla u^N] \cdot [\nabla v^N] \, ds \\ &= \sum_{e \in \mathcal{E}_x} \gamma N^{-2} \int_e [u_x^N][v_x^N] \, ds + \sum_{e \in \mathcal{E}_y} \gamma N^{-2} \int_e [u_y^N][v_y^N] \, ds \\ &=: J_x(u^N, v^N) + J_y(u^N, v^N). \end{aligned}$$

Lemma 6.60. *Let $u^I \in S^N$ be the piecewise bilinear nodal interpolant of the solution u to (6.48). Then there exists a constant C, which is independent of ε, N, and v^N, such that*

$$|J(u-u^I, v^N)| \le CN^{-7/4}\|v^N\|_{CIP} \quad \forall v^N \in V^N.$$

Proof. From above, $J(u-u^I, v^N) = J_x(u-u^I, v^N) + J_y(u-u^I, v^N)$. By symmetry we need analyse only the term $J_x(u-u^I, v^N)$. Recalling the decomposition of u in (4.14), one has

$$\begin{aligned} J_x(u-u^I, v^N) &= J_x(S-S^I, v^N) + J_x(E_2-E_2^I, v^N) \\ &\quad + J_x(E_1+E_{12}-E_1^I-E_{12}^I, v^N) \\ &=: T_0 + T_1 + T_2. \end{aligned} \tag{6.55}$$

Here S^1 is the nodal interpolant of S, E_2^I is the nodal interpolant of E_2, etc.

Now

$$\begin{aligned} T_0 &= \gamma N^{-2} \sum_{i=1}^{N/2-1} \sum_{j=0}^{N/2-1} \int_{e_{j;i}} [(S-S^I)_x]\,[v_x^N]\,ds \\ &= \gamma N^{-2} \sum_{j=0}^{N/2-1} \sum_{i=1}^{N/2-1} \int_{e_{j;i}} \left(-S_x^I|_{\tau_{i,j}} + S_x^I|_{\tau_{i-1,j}}\right) \cdot \left(v_x^N|_{\tau_{i,j}} - v_x^N|_{\tau_{i-1,j}}\right) ds \\ &= \gamma N^{-2} \sum_{j=0}^{N/2-1} \int_{y_j}^{y_{j+1}} \left([-S_x^I]|_{e_{j;N/2-1}} \cdot v_x^N|_{\tau_{N/2-1,j}} + [S_x^I]|_{e_{j;1}} \cdot v_x^N|_{\tau_{0,j}}\right) dy \\ &\quad + \gamma N^{-2} \sum_{j=0}^{N/2-1} \sum_{i=1}^{N/2-2} \int_{y_j}^{y_{j+1}} \left(S_x^I|_{\tau_{i-1,j}} - 2S_x^I|_{\tau_{i,j}} + S_x^I|_{\tau_{i+1,j}}\right) \cdot v_x^N|_{\tau_{i,j}}\,dy \\ &=: I_1 + I_2. \end{aligned}$$

As $v_x^N|_{\tau_{i,j}}$ is a function of y only for each i and j, one has

$$\int_{y_j}^{y_{j+1}} |v_x^N|\,dy = \frac{1}{x_{i+1}-x_i} \int_{x_i}^{x_{i+1}} \int_{y_j}^{y_{j+1}} |v_x^N|\,dx\,dy. \tag{6.56}$$

Set $\tilde{\Omega}_{11} := \sum_{j=0}^{N/2-1}(\tau_{N/2-1,j} \cup \tau_{0,j})$. The smoothness of S and (6.56) yield

$$\begin{aligned}
|I_1| &\leq CN^{-3} \sum_{j=0}^{N/2-1} \left(\int_{y_j}^{y_{j+1}} |(v_x^N|_{\tau_{N/2-1,j}})|\, dy + \int_{y_j}^{y_{j+1}} |(v_x^N|_{\tau_{0,j}})|\, dy \right) \\
&\leq CN^{-2} \sum_{j=0}^{N/2-1} \left(\|v_x^N\|_{L^1(\tau_{N/2-1,j})} + \|v_x^N\|_{L^1(\tau_{0,j})} \right) \\
&\leq CN^{-2} \cdot N^{-1/2} \|v_x^N\|_{\tilde{\Omega}_{11}} \\
&\leq CN^{-7/4} \|v^N\|_{CIP},
\end{aligned} \tag{6.57}$$

where we used a Cauchy–Schwarz inequality and $\text{meas}(\tilde{\Omega}_{11}) \leq CN^{-1}$, then invoked Lemma 6.59. Similarly, one has

$$\begin{aligned}
|I_2| &\leq CN^{-4} \sum_{j=0}^{N/2-1} \sum_{i=1}^{N/2-2} \int_{y_j}^{y_{j+1}} \left|(v_x^N|_{\tau_{i,j}})\right| \, dy \\
&\leq CN^{-3} \sum_{j=0}^{N/2-1} \sum_{i=1}^{N/2-2} \int_{x_i}^{x_{i+1}} \int_{y_j}^{y_{j+1}} \left|v_x^N\right| \, dx\, dy \\
&\leq CN^{-3} \|v_x^N\|_{L^1(\Omega_{11})} \\
&\leq CN^{-2} \|v^N\|_{L^1(\Omega_{11})} \\
&\leq CN^{-2} \|v^N\|_{CIP},
\end{aligned} \tag{6.58}$$

where an inverse estimate was used. From (6.57) and (6.58) we obtain

$$|T_0| \leq CN^{-7/4} \|v^N\|_{CIP}. \tag{6.59}$$

The x-derivatives of E_2 are bounded independently of ε by (4.14), so we can analyse T_1 similarly to T_0 and get

$$|T_1| \leq CN^{-5/2} \|v^N\|_{CIP} \tag{6.60}$$

since $m \geq 5/2$ in our choice of the Shishkin mesh transition points.

Next, inverse estimates and (4.14) yield

$$\begin{aligned}
\left\|[(E_1^I + E_{12}^I)_x]\right\|_{L^\infty(e_{j;i})} &\leq C\|(E_1^I + E_{12}^I)_x\|_{L^\infty(\tau_{i-1,j} \cup \tau_{i,j})} \\
&\leq CN\|E_1^I + E_{12}^I\|_{L^\infty(\tau_{i-1,j} \cup \tau_{i,j})} \\
&\leq CN^{-3/2},
\end{aligned}$$

again because $m \geq 5/2$. This bound and (6.56) give us

$$
\begin{aligned}
|T_2| &\leq \gamma N^{-2} \sum_{i=1}^{N/2-1} \sum_{j=0}^{N/2-1} \|[(E_1^I + E_{12}^I)_x]\|_{L^\infty(e_{j;i})} \|[v_x^N]\|_{L^1(e_{j;i})} \\
&\leq CN^{-2} \sum_{i=1}^{N/2-1} \sum_{j=0}^{N/2-1} N^{-3/2} \cdot N \|v_x^N\|_{L^1(\tau_{i-1,j} \cup \tau_{i,j})} \qquad (6.61) \\
&\leq CN^{-5/2} \|v_x^N\|_{L^1(\Omega_{11})} \\
&\leq CN^{-2} \cdot N^{-1/2} \|v_x^N\|_{\Omega_{11}} \\
&\leq CN^{-7/4} \|v^N\|_{CIP}
\end{aligned}
$$

by a Cauchy–Schwarz inequality and Lemma 6.59.

Substituting (6.59)–(6.61) into (6.55), we obtain $|J_x(u - u^I, v^N)| \leq CN^{-7/4}\|v^N\|_{CIP}$, as desired. □

The main convergence result for our CIP can now be derived.

Theorem 6.61. *Let $u^I \in S^N$ be the bilinear interpolant to the solution u of* (6.48), *and let $u^N \in S^N$ be the CIP solution of* (6.50). *Then there exists a constant C such that*

$$\|u^I - u^N\|_{CIP} \leq C(N^{-7/4} + \varepsilon N^{-3/2}).$$

Proof. Coercivity (6.52) and Galerkin orthogonality (6.53) yield

$$
\begin{aligned}
\|u^I - u^N\|_{CIP}^2 &\leq a(u^I - u^N, u^I - u^N) \\
&= a(u^I - u, u^I - u^N) \\
&= B(u^I - u, u^I - u^N) + J(u^I - u, u^I - u^N).
\end{aligned}
$$

In [**Zha03**] it is proved that

$$|B(u - u^I, v^N)| \leq C(N^{-2}\ln^2 N + \varepsilon N^{-3/2})\|v^N\|_{CIP} \quad \text{for all } v^N \in S^N.$$

Taking $v^N = u^I - u^N$ in this inequality and in Lemma 6.60, the result follows. □

From Theorem 6.61 it is easy to obtain an estimate for the error $u - u^N$.

Corollary 6.62. *Let $u^N \in S^N$ be the solution of* (6.50). *Then there exists a constant C such that*

$$\|u - u^N\|_0 \leq C\left[N^{-7/4} + \varepsilon N^{-3/2} + \varepsilon^{1/2}(N^{-1}\ln N)^2\right].$$

Proof. Theorem 6.61 clearly implies that $\|u^I - u^N\|_0 \leq C(N^{-7/4} + \varepsilon N^{-3/2})$. But $\|u - u_0^I\| \leq C[N^{-2} + \varepsilon^{1/2}(N^{-1}\ln N)^2]$ by [**RST08**, Lemma 3.107]. Use a triangle inequality to finish the proof. □

Remark 6.63 (Supercloseness and postprocessing). The definition of the CIP norm implies that

$$\|u - u^N\|_{CIP} \geq \left[\varepsilon|u - u^N|_1^2 + C_5\|u - u^N\|^2\right]^{1/2}.$$

But it is well known that on our Shishkin mesh the bound

$$\left[\varepsilon|u - u^N|_1^2 + C_5\|u - u^N\|^2\right]^{1/2} \leq CN^{-1}\ln N \tag{6.62}$$

of [**RST08**, Lemma 3.107] is *sharp*. Consequently, $\|u - u^N\|_{CIP} \leq CN^{-1}\ln N$ is the best possible bound, which implies that the bound of Theorem 6.61 is a *supercloseness* result, i.e., the computed solution u^N approximates a certain finite-dimensional interpolant of the true solution more accurately (i.e., with a higher order of convergence) than it approximates the solution u itself. Recall Remark 6.51. As described in that remark, one can exploit the supercloseness property by postprocessing the solution in a simple procedure that produces a piecewise quadratic solution Pu^N on our mesh with the property

$$\left(\varepsilon|u - Pu^N|_1^2 + \|u - Pu^N\|_0^2\right)^{1/2} \leq C(\varepsilon N^{-3/2} + N^{-7/4}).$$

In this inequality we are measuring the error between Pu^N and the true solution u, which is more satisfactory than Theorem 6.61 where the error between u^N and the interpolant u^I was estimated.

Exercise 6.64. Prove that the bound (6.62) is sharp by considering examples of solutions u.

In the interesting paper [**Sch08**], Schieweck discusses and analyses the effect of imposing boundary conditions weakly in the CIP method, using a variation of the approach of Nitsche (Example 6.52).

6.9. Adaptive methods

Adaptive FEMs compute a solution to a boundary value problem on some conventional (e.g., equidistant) mesh using some stable method such as SDFEM, then use this solution to compute a posteriori some local error estimator that gives guidance on where one should refine or coarsen the mesh to obtain a mesh better suited to the boundary value problem. On this new mesh one then computes a fresh solution to the problem, then the mesh is again modified based on the local error estimator. The process is continued iteratively until some stopping criterion is reached. See [**AO00**] or [**BS08**, Chap. 9] for a more precise description.

There is perhaps a general consensus that in the long run adaptive methods will provide the most satisfactory approach to solving convection-diffusion problems, but today their behaviour when applied to such problems is still poorly understood, despite many published numerical experiments.

John [**Joh00**] gives numerical examples of how apparently reasonable error estimators can yield inaccurate solutions to convection-diffusion problems.

A difficulty with the theory of a posteriori error estimators for convection-diffusion problems is that published inequalities relating the estimator to the true error frequently contain multiplicative factors that depend badly on the small diffusion parameter ε. This seriously undermines the validity of the estimator. Below we shall confine our discussion to a few ε-independent results that have been obtained.

Remark 6.65 (Adaptive finite difference methods)**.** For the one-dimensional problem (2.14), an adaptive-mesh algorithm that is based on arc-length equidistribution (where mesh points are moved but no points are created or deleted) is analysed by [**KS01**], using earlier a posteriori bounds from [**Kop01**]. It is shown that, starting from an equidistant mesh with N subintervals, after $\mathcal{O}(\ln(1/\varepsilon)/(\ln N))$ iterations, one obtains a computed solution u^N that resolves the layer with, moreover, $|u(x_i) - u_i^N| \le CN^{-1}$ for all i. The underlying numerical method is simple upwinding so this is a finite difference approach, but we include it here since it is a clear convergence result for an adaptive method and few such results exist for convection-diffusion problems. It seems difficult to extend this type of result to two-dimensional problems. A more leisurely and readable exposition of this material is given in [**Kop07a**]. Adaptive methods for finite difference methods applied to convection diffusion have been examined by Kopteva et al. [**FK11, LK10**].

In [**San01**] the *residual-free bubble* FEM is considered; this method is related to the SDFEM [**BMS00**]. An error estimator based on element residuals and jumps in the normal derivative of the solution across edges is shown to be robust for (4.1); i.e., the global value of the estimator is equivalent to the true error up to a constant factor that is independent of ε, but the norm in which the true error is measured is

$$w \mapsto \varepsilon|w|_{H^1(\Omega)} + \|\mathbf{a}\cdot\nabla w\|_{H^{-1}(\Omega)},$$

which is weak—the factor multiplying $|\cdot|_{H^1(\Omega)}$ is ε, not the more natural $\varepsilon^{1/2}$ that appears in the weighted energy norm $\|\cdot\|_{1,\varepsilon}$ of section 4.2.

The *dual-weighted-residual* method for goal-oriented error estimation has been successfully applied to convection-diffusion problems by various authors; see [**EEHJ96**] and [**BR03**]. Here the aim is to adapt the mesh in order to compute accurately some functional of the solution but not the solution itself. See the survey paper [**GS02**].

Verfürth [**Ver05**] shows that for SDFEM the error in the computed solution is equivalent (up to a constant factor that is independent of ε) to the global value of each of three different estimators (one based on element and edge residuals, one based on the solution of local Dirichlet problems, one based on the solution of local Neumann problems). The true error is measured in a norm

$$w \mapsto \|w\|_{1,\varepsilon} + \|w\|_*,$$

where $\|\cdot\|$ is the dual norm on $H^{-1}(\Omega)$ defined by

$$\|w\|_* = \sup_{v \in H_0^1(\Omega)\setminus\{0\}} \frac{(w,v)}{\|v\|_{1,\varepsilon}},$$

with $(\cdot,\cdot)$ the corresponding duality pairing. (This special norm is used to bound the convective term.) But the paper assumes that the mesh is quasi-uniform, which excludes the long thin elements that one expects an adaptive code to construct when solving a convection-diffusion problem.

See also [**ANS11**, **EV11**, **ZS11**] and their references. An excellent survey of the state of the art in adaptive finite element methods for convection-diffusion problems is given in [**JKN18**].

Still, despite much research activity in this area, we do not have today a satisfactory adaptive method for two-dimensional convection-diffusion problems that, starting from an ordinary coarse mesh, is guaranteed to produce a layer-adapted mesh with a bound on the error in the computed solution in some reasonably strong norm.

Remark 6.66 (Adaptivity for reaction-diffusion problems)**.** For singularly perturbed reaction-diffusion problems, see [**Kop17**], which gives a good overview of the current state of research on this topic.

Chapter 7

Concluding Remarks

Our survey of numerical methods for steady-state convection-diffusion problems has not been exhaustive. For example, to learn about *hp* finite element methods, see [**EM07**, **HSS02**, **Sch98**, **ZS11**] and also [**Lin10**, **Mel02**], where singularly perturbed linear reaction-diffusion problems are examined. Subgrid modelling is examined in [**AEFES09**, **RST08**].

Several numerical methods are compared in [**LS01b**] where a two-dimensional Shishkin mesh is used to solve a problem on the unit square whose solution has exponential boundary layers and a corner layer. An interesting and detailed numerical comparison of several finite element methods for the challenging Hemker problem of section 4.2.2 is presented in [**ACF+11**].

The numerical analysis and solution of convection-diffusion problems on polygonal regions, where the solution is assumed to exhibit boundary but not interior layers and one has sufficient compatibility of the data at the corners of the domain, is by now fairly well understood in the framework of Shishkin meshes combined with finite difference or finite element methods. When we consider interior layers (and the effects of data incompatibilities at corners) our grasp is much less, sure and there are several competing methods. In the long run, our view is that adaptive methods will triumph over all types of convection-diffusion problems, but much work remains to be done.

For general surveys of methods for convection-diffusion problems see [**CGL11**, **FR11**, **HKOS09**, **Lin10**, **Mor96**, **Roo12**, **RST08**] and the very readable paper [**JKN18**].

Bibliography

[ABCM02] D. N. Arnold, F. Brezzi, B. Cockburn, and L. D. Marini, *Unified analysis of discontinuous Galerkin methods for elliptic problems*, SIAM J. Numer. Anal. **39** (2001/02), no. 5, 1749–1779, DOI 10.1137/S0036142901384162. MR1885715

[ACF+11] M. Augustin, A. Caiazzo, A. Fiebach, J. Fuhrmann, V. John, A. Linke, and R. Umla, *An assessment of discretizations for convection-dominated convection-diffusion equations*, Comput. Methods Appl. Mech. Engrg. **200** (2011), no. 47-48, 3395–3409, DOI 10.1016/j.cma.2011.08.012. MR2844065

[AD92] T. Apel and M. Dobrowolski, *Anisotropic interpolation with applications to the finite element method* (English, with German summary), Computing **47** (1992), no. 3-4, 277–293, DOI 10.1007/BF02320197. MR1155498

[AEFES09] B. Achchab, M. El Fatini, A. Ern, and A. Souissi, *A posteriori error estimates for subgrid viscosity stabilized approximations of convection-diffusion equations*, Appl. Math. Lett. **22** (2009), no. 9, 1418–1424, DOI 10.1016/j.aml.2008.12.006. MR2536825

[AK90] O. Axelsson and L. Kolotilina, *Monotonicity and discretization error estimates*, SIAM J. Numer. Anal. **27** (1990), no. 6, 1591–1611, DOI 10.1137/0727093. MR1080340

[AK96] V. B. Andreev and N. V. Kopteva, *Investigation of difference schemes with an approximation of the first derivative by a central difference relation* (Russian, with Russian summary), Zh. Vychisl. Mat. i Mat. Fiz. **36** (1996), no. 8, 101–117; English transl., Comput. Math. Math. Phys. **36** (1996), no. 8, 1065–1078 (1997). MR1407730

[And02] V. B. Andreev, *Pointwise and weighted a priori estimates for the solution and its first derivative of a singularly perturbed convection-diffusion equation* (Russian, with Russian summary), Differ. Uravn. **38** (2002), no. 7, 918–929, 1005, DOI 10.1023/A:1021103512850; English transl., Differ. Equ. **38** (2002), no. 7, 972–984. MR2005755

[And06] V. B. Andreev, *On the accuracy of grid approximations of nonsmooth solutions of a singularly perturbed reaction-diffusion equation in the square* (Russian, with Russian summary), Differ. Uravn. **42** (2006), no. 7, 895–906, 1005, DOI 10.1134/S0012266106070044; English transl., Differ. Equ. **42** (2006), no. 7, 954–966. MR2294140

[ANS11] T. Apel, S. Nicaise, and D. Sirch, *A posteriori error estimation of residual type for anisotropic diffusion-convection-reaction problems*, J. Comput. Appl. Math. **235** (2011), no. 8, 2805–2820, DOI 10.1016/j.cam.2010.11.032. MR2763187

[AO00] M. Ainsworth and J. T. Oden, *A posteriori error estimation in finite element analysis*, Pure and Applied Mathematics (New York), Wiley-Interscience [John Wiley & Sons], New York, 2000. MR1885308

[Ape99] T. Apel, *Anisotropic finite elements: local estimates and applications*, Advances in Numerical Mathematics, B. G. Teubner, Stuttgart, 1999. MR1716824

[AS55] D. N. d. G. Allen and R. V. Southwell, *Relaxation methods applied to determine the motion, in two dimensions, of a viscous fluid past a fixed cylinder*, Quart. J. Mech. Appl. Math. **8** (1955), 129–145, DOI 10.1093/qjmam/8.2.129. MR0070367

[AT04] J. E. Akin and T. E. Tezduyar, *Calculation of the advective limit of the supg stabilization parameter for linear and higher-order elements*, Comput. Methods Appl. Mech. Engrg. **193** (2004), 1909–1922.

[BE06] E. Burman and A. Ern, *Continuous interior penalty hp-finite element methods for transport operators*, Numerical mathematics and advanced applications, Springer, Berlin, 2006, pp. 504–511, DOI 10.1007/978-3-540-34288-5_46. MR2303678

[BGL09] E. Burman, J. Guzmán, and D. Leykekhman, *Weighted error estimates of the continuous interior penalty method for singularly perturbed problems*, IMA J. Numer. Anal. **29** (2009), no. 2, 284–314, DOI 10.1093/imanum/drn001. MR2491428

[BH04] E. Burman and P. Hansbo, *Edge stabilization for Galerkin approximations of convection-diffusion-reaction problems*, Comput. Methods Appl. Mech. Engrg. **193** (2004), no. 15-16, 1437–1453, DOI 10.1016/j.cma.2003.12.032. MR2068903

[BMS00] F. Brezzi, D. Marini, and E. Süli, *Residual-free bubbles for advection-diffusion problems: the general error analysis*, Numer. Math. **85** (2000), no. 1, 31–47, DOI 10.1007/s002110050476. MR1751366

[Boh81] E. Bohl, *Finite Modelle gewöhnlicher Randwertaufgaben* (German), Leitfäden der Angewandten Mathematik und Mechanik [Guides to Applied Mathematics and Mechanics], vol. 51, B. G. Teubner, Stuttgart, 1981. Teubner Studienbücher: Mathematik. [Teubner Study Books: Mathematics]. MR633643

[Boy99] J. P. Boyd, *The devil's invention: asymptotic, superasymptotic and hyperasymptotic series*, Acta Appl. Math. **56** (1999), no. 1, 1–98, DOI 10.1023/A:1006145903624. MR1698036

[BPP15] P. Bochev, M. Perego, and K. Peterson, *Formulation and analysis of a parameter-free stabilized finite element method*, SIAM J. Numer. Anal. **53** (2015), no. 5, 2363–2388, DOI 10.1137/14096284X. MR3414468

[BR84] K. Bohmer and R. Rannacher, *Defect correction methods: Theory and applications*, Springer-Verlag, Berlin, 1984.

[BR94] F. Brezzi and A. Russo, *Choosing bubbles for advection-diffusion problems*, Math. Models Methods Appl. Sci. **4** (1994), no. 4, 571–587, DOI 10.1142/S0218202594000327. MR1291139

[BR03] W. Bangerth and R. Rannacher, *Adaptive finite element methods for differential equations*, Lectures in Mathematics ETH Zürich, Birkhäuser Verlag, Basel, 2003. MR1960405

[BS08] S. C. Brenner and L. R. Scott, *The mathematical theory of finite element methods*, 3rd ed., Texts in Applied Mathematics, vol. 15, Springer, New York, 2008. MR2373954

[BS16] E. Burman and F. Schieweck, *Local CIP stabilization for composite finite elements*, SIAM J. Numer. Anal. **54** (2016), no. 3, 1967–1992, DOI 10.1137/15M1039390. MR3516865

[Bur05] E. Burman, *A unified analysis for conforming and nonconforming stabilized finite element methods using interior penalty*, SIAM J. Numer. Anal. **43** (2005), no. 5, 2012–2033, DOI 10.1137/S0036142903437374. MR2192329

[Bur12] E. Burman, *A penalty-free nonsymmetric Nitsche-type method for the weak imposition of boundary conditions*, SIAM J. Numer. Anal. **50** (2012), no. 4, 1959–1981, DOI 10.1137/10081784X. MR3022206

[BY91] A. Brandt and I. Yavneh, *Inadequacy of first-order upwind difference schemes for some recirculating flows*, J. Comput. Phys. **93** (1991), no. 1, 128–143, DOI 10.1016/0021-9991(91)90076-W. MR1097117

[CGL11] C. Clavero, J. L. Gracia, and F. J. Lisbona (eds.), *BAIL 2010—boundary and interior layers, computational and asymptotic methods*, Lecture Notes in Computational Science and Engineering, vol. 81, Springer, Heidelberg, 2011. Selected papers from the conference held at the University of Zaragoza, Zaragoza, July 5–9, 2010. MR2849747

[CGO05] C. Clavero, J. L. Gracia, and E. O'Riordan, *A parameter robust numerical method for a two dimensional reaction-diffusion problem*, Math. Comp. **74** (2005), no. 252, 1743–1758, DOI 10.1090/S0025-5718-05-01762-X. MR2164094

[CKSL+14] Q. Cai, S. Kollmannsberger, E. Sala-Lardies, A. Huerta, and E. Rank, *On the natural stabilization of convection dominated problems using high order Bubnov-Galerkin finite elements*, Comput. Math. Appl. **66** (2014), no. 12, 2545–2558, DOI 10.1016/j.camwa.2013.09.009. MR3128578

[Cod11] R. Codina, *Finite element approximation of the convection-diffusion equation: subgrid-scale spaces, local instabilities and anisotropic space-time discretizations*, BAIL 2010—boundary and interior layers, computational and asymptotic methods, Lect. Notes Comput. Sci. Eng., vol. 81, Springer, Heidelberg, 2011, pp. 85–97, DOI 10.1007/978-3-642-19665-2_10. MR2849731

[CX08] L. Chen and J. Xu, *Stability and accuracy of adapted finite element methods for singularly perturbed problems*, Numer. Math. **109** (2008), no. 2, 167–191, DOI 10.1007/s00211-007-0118-6. MR2385650

[DD76] J. Douglas Jr. and T. Dupont, *Interior penalty procedures for elliptic and parabolic Galerkin methods*, Computing methods in applied sciences (Second Internat. Sympos., Versailles, 1975), Springer, Berlin, 1976, pp. 207–216. Lecture Notes in Phys., Vol. 58. MR0440955

[DG11] L. Demkowicz and J. Gopalakrishnan, *A class of discontinuous Petrov-Galerkin methods. II. Optimal test functions*, Numer. Methods Partial Differential Equations **27** (2011), no. 1, 70–105, DOI 10.1002/num.20640. MR2743600

[DLP12] R. G. Durán, A. L. Lombardi, and M. I. Prieto, *Superconvergence for finite element approximation of a convection-diffusion equation using graded meshes*, IMA J. Numer. Anal. **32** (2012), no. 2, 511–533, DOI 10.1093/imanum/drr005. MR2911398

[Dör99] W. Dörfler, *Uniform a priori estimates for singularly perturbed elliptic equations in multidimensions*, SIAM J. Numer. Anal. **36** (1999), no. 6, 1878–1900, DOI 10.1137/S0036142998341325. MR1712153

[DPE12] D. A. Di Pietro and A. Ern, *Mathematical aspects of discontinuous Galerkin methods*, Mathématiques & Applications (Berlin) [Mathematics & Applications], vol. 69, Springer, Heidelberg, 2012. MR2882148

[DR97] M. Dobrowolski and H.-G. Roos, *A priori estimates for the solution of convection-diffusion problems and interpolation on Shishkin meshes*, Z. Anal. Anwendungen **16** (1997), no. 4, 1001–1012, DOI 10.4171/ZAA/801. MR1615644

[EEHJ96] K. Eriksson, D. Estep, P. Hansbo, and C. Johnson, *Computational differential equations*, Cambridge University Press, Cambridge, 1996. MR1414897

[EM07] T. Eibner and J. M. Melenk, *An adaptive strategy for hp-FEM based on testing for analyticity*, Comput. Mech. **39** (2007), no. 5, 575–595, DOI 10.1007/s00466-006-0107-0. MR2288643

[EV11] A. Ern and M. Vohralík, *A unified framework for a posteriori error estimation in elliptic and parabolic problems with application to finite volumes*, Finite volumes for complex applications VI. Problems & perspectives. Volume 1, 2, Springer Proc. Math., vol. 4, Springer, Heidelberg, 2011, pp. 821–837, DOI 10.1007/978-3-642-20671-9_85. MR2882361

[Eva10] L. C. Evans, *Partial differential equations*, 2nd ed., Graduate Studies in Mathematics, vol. 19, American Mathematical Society, Providence, RI, 2010. MR2597943

[FHM+00] P. A. Farrell, A. F. Hegarty, J. J. H. Miller, E. O'Riordan, and G. I. Shishkin, *Robust computational techniques for boundary layers*, Applied Mathematics (Boca Raton), vol. 16, Chapman & Hall/CRC, Boca Raton, FL, 2000. MR1750671

[FK11] S. Franz and N. Kopteva, *Green's function estimates for a singularly perturbed convection-diffusion problem in three dimensions*, Int. J. Numer. Anal. Model. Ser. B **2** (2011), no. 2-3, 124–141. MR2837411

[FK12] S. Franz and N. Kopteva, *Green's function estimates for a singularly perturbed convection-diffusion problem*, J. Differential Equations **252** (2012), no. 2, 1521–1545, DOI 10.1016/j.jde.2011.07.033. MR2853549

[FR11] S. Franz and H.-G. Roos, *The capriciousness of numerical methods for singular perturbations*, SIAM Rev. **53** (2011), no. 1, 157–173, DOI 10.1137/090757344. MR2785883

[Fra08] S. Franz, *Continuous interior penalty method on a Shishkin mesh for convection-diffusion problems with characteristic boundary layers*, Comput. Methods Appl. Mech. Engrg. **197** (2008), no. 45-48, 3679–3686, DOI 10.1016/j.cma.2008.02.019. MR2458107

[FRSW99] B. Fischer, A. Ramage, D. J. Silvester, and A. J. Wathen, *On parameter choice and iterative convergence for stabilised discretisations of advection-diffusion problems*, Comput. Methods Appl. Mech. Engrg. **179** (1999), no. 1-2, 179–195, DOI 10.1016/S0045-7825(99)00037-7. MR1716843

[GFL+83] H. Goering, A. Felgenhauer, G. Lube, H.-G. Roos, and L. Tobiska, *Singularly perturbed differential equations*, Mathematical Research, vol. 13, Akademie-Verlag, Berlin, 1983. MR718115

[GK03] J. Gopalakrishnan and G. Kanschat, *A multilevel discontinuous Galerkin method*, Numer. Math. **95** (2003), no. 3, 527–550, DOI 10.1007/s002110200392. MR2012931

[Gos13] L. Gosse, *Computing qualitatively correct approximations of balance laws: Exponential-fit, well-balanced and asymptotic-preserving*, SIMAI Springer Series, vol. 2, Springer, Milan, 2013. MR3053000

[Gri85] P. Grisvard, *Elliptic problems in nonsmooth domains*, Monographs and Studies in Mathematics, vol. 24, Pitman (Advanced Publishing Program), Boston, MA, 1985. MR775683

[GS02] M. B. Giles and E. Süli, *Adjoint methods for PDEs: a posteriori error analysis and postprocessing by duality*, Acta Numer. **11** (2002), 145–236, DOI 10.1017/S096249290200003X. MR2009374

[GT01] D. Gilbarg and N. S. Trudinger, *Elliptic partial differential equations of second order*, Classics in Mathematics, Springer-Verlag, Berlin, 2001. Reprint of the 1998 edition. MR1814364

[GT17] S. Ganesan and L. Tobiska, *Finite elements: Theory and algorithms*, Cambridge-IISc Series, Cambridge University Press, Cambridge, 2017. MR3752652

[HB79] T. J. R. Hughes and A. Brooks, *A multidimensional upwind scheme with no crosswind diffusion*, Finite element methods for convection dominated flows (Papers, Winter Ann. Meeting Amer. Soc. Mech. Engrs., New York, 1979), AMD, vol. 34, Amer. Soc. Mech. Engrs. (ASME), New York, 1979, pp. 19–35. MR571681

[Hem96] P. W. Hemker, *A singularly perturbed model problem for numerical computation*, J. Comput. Appl. Math. **76** (1996), no. 1-2, 277–285, DOI 10.1016/S0377-0427(96)00113-6. MR1423523

[HH14] H. Han and Z. Huang, *The tailored finite point method*, Comput. Methods Appl. Math. **14** (2014), no. 3, 321–345, DOI 10.1515/cmam-2014-0012. MR3228914

[HK90] H. Han and R. B. Kellogg, *Differentiability properties of solutions of the equation* $-\epsilon^2\Delta u + ru = f(x, y)$ *in a square*, SIAM J. Math. Anal. **21** (1990), no. 2, 394–408, DOI 10.1137/0521022. MR1038899

[HKOS09] A. F. Hegarty, N. Kopteva, E. O'Riordan, and M. Stynes (eds.), *BAIL 2008—boundary and interior layers*, Lecture Notes in Computational Science and Engineering, vol. 69, Springer-Verlag, Berlin, 2009. MR2547533

[HS01] P. Houston and E. Süli, *Stabilised hp-finite element approximation of partial differential equations with nonnegative characteristic form*, Computing **66** (2001), no. 2, 99–119, DOI 10.1007/s006070170030. Archives for scientific computing. Numerical methods for transport-dominated and related problems (Magdeburg, 1999). MR1825801

[HSS02] P. Houston, C. Schwab, and E. Süli, *Discontinuous hp-finite element methods for advection-diffusion-reaction problems*, SIAM J. Numer. Anal. **39** (2002), no. 6, 2133–2163, DOI 10.1137/S0036142900374111. MR1897953

[Il′69] A. M. Il′in, *A difference scheme for a differential equation with a small parameter multiplying the highest derivative* (Russian), Mat. Zametki **6** (1969), 237–248. MR0260195

[Il′92] A. M. Il′in, *Matching of asymptotic expansions of solutions of boundary value problems*, Translations of Mathematical Monographs, vol. 102, American Mathematical Society, Providence, RI, 1992. Translated from the Russian by V. Minachin [V. V. Minakhin]. MR1182791

[JKN18] Volker John, Petr Knobloch, and Julia Novo, *Finite elements for scalar convection-dominated equations and incompressible flow problems—a never ending story?*, Comput. Vis. Sci. (2018), To appear.

[JKS11] V. John, P. Knobloch, and S. B. Savescu, *A posteriori optimization of parameters in stabilized methods for convection-diffusion problems—Part I*, Comput. Methods Appl. Mech. Engrg. **200** (2011), no. 41-44, 2916–2929, DOI 10.1016/j.cma.2011.04.016. MR2824164

[JN13] V. John and J. Novo, *A robust SUPG norm a posteriori error estimator for stationary convection-diffusion equations*, Comput. Methods Appl. Mech. Engrg. **255** (2013), 289–305, DOI 10.1016/j.cma.2012.11.019. MR3029040

[Joh00] V. John, *A numerical study of a posteriori error estimators for convection-diffusion equations*, Comput. Methods Appl. Mech. Engrg. **190** (2000), no. 5-7, 757–781, DOI 10.1016/S0045-7825(99)00440-5. MR1800575

[KC81] J. Kevorkian and J. D. Cole, *Perturbation methods in applied mathematics*, Applied Mathematical Sciences, vol. 34, Springer-Verlag, New York-Berlin, 1981. MR608029

[KC96] J. Kevorkian and J. D. Cole, *Multiple scale and singular perturbation methods*, Applied Mathematical Sciences, vol. 114, Springer-Verlag, New York, 1996. MR1392475

[Kev00] J. Kevorkian, *Partial differential equations*, 2nd ed., Texts in Applied Mathematics, vol. 35, Springer-Verlag, New York, 2000. Analytical solution techniques. MR1728947

[KLR02] T. Knopp, G. Lube, and G. Rapin, *Stabilized finite element methods with shock capturing for advection-diffusion problems*, Comput. Methods Appl. Mech. Engrg. **191** (2002), no. 27-28, 2997–3013, DOI 10.1016/S0045-7825(02)00222-0. MR1903196

[KLS08] R. B. Kellogg, T. Linß, and M. Stynes, *A finite difference method on layer-adapted meshes for an elliptic reaction-diffusion system in two dimensions*, Math. Comp. **77** (2008), no. 264, 2085–2096, DOI 10.1090/S0025-5718-08-02125-X. MR2429875

[KO10] N. Kopteva and E. O'Riordan, *Shishkin meshes in the numerical solution of singularly perturbed differential equations*, Int. J. Numer. Anal. Model. **7** (2010), no. 3, 393–415. MR2644280

[Kop01] N. Kopteva, *Maximum norm a posteriori error estimates for a one-dimensional convection-diffusion problem*, SIAM J. Numer. Anal. **39** (2001), no. 2, 423–441, DOI 10.1137/S0036142900368642. MR1860270

[Kop03] N. Kopteva, *Error expansion for an upwind scheme applied to a two-dimensional convection-diffusion problem*, SIAM J. Numer. Anal. **41** (2003), no. 5, 1851–1869, DOI 10.1137/S003614290241074X. MR2035009

[Kop04] N. Kopteva, *How accurate is the streamline-diffusion FEM inside characteristic (boundary and interior) layers?*, Comput. Methods Appl. Mech. Engrg. **193** (2004), no. 45-47, 4875–4889, DOI 10.1016/j.cma.2004.05.008. MR2097760

[Kop07a] N. Kopteva, *Convergence theory of moving grid methods*, Adaptive Computations: Theory and Algorithms (Tao Tang and Jinchao Xu, eds.), Science Press, Beijing, 2007, pp. 147–191.

[Kop07b] N. Kopteva, *Maximum norm error analysis of a 2D singularly perturbed semilinear reaction-diffusion problem*, Math. Comp. **76** (2007), no. 258, 631–646, DOI 10.1090/S0025-5718-06-01938-7. MR2291831

[Kop14] N. Kopteva, *Linear finite elements may be only first-order pointwise accurate on anisotropic triangulations*, Math. Comp. **83** (2014), no. 289, 2061–2070, DOI 10.1090/S0025-5718-2014-02820-2. MR3223324

[Kop17] N. Kopteva, *Fully computable a posteriori error estimator using anisotropic flux equilibration on anisotropic meshes*, ArXiv e-prints (2017).

[KS01] N. Kopteva and M. Stynes, *A robust adaptive method for a quasi-linear one-dimensional convection-diffusion problem*, SIAM J. Numer. Anal. **39** (2001), no. 4, 1446–1467, DOI 10.1137/S003614290138471X. MR1870850

[KS05] R. B. Kellogg and M. Stynes, *Corner singularities and boundary layers in a simple convection-diffusion problem*, J. Differential Equations **213** (2005), no. 1, 81–120, DOI 10.1016/j.jde.2005.02.011. MR2139339

[KS06] R. B. Kellogg and M. Stynes, *A singularly perturbed convection-diffusion problem in a half-plane*, Appl. Anal. **85** (2006), no. 12, 1471–1485, DOI 10.1080/00036810601066574. MR2282997

[KS07] R. B. Kellogg and M. Stynes, *Sharpened bounds for corner singularities and boundary layers in a simple convection-diffusion problem*, Appl. Math. Lett. **20** (2007), no. 5, 539–544, DOI 10.1016/j.aml.2006.08.001. MR2303990

[KT78] R. B. Kellogg and A. Tsan, *Analysis of some difference approximations for a singular perturbation problem without turning points*, Math. Comp. **32** (1978), no. 144, 1025–1039, DOI 10.2307/2006331. MR0483484

[KT11] P. Knobloch and L. Tobiska, *On the stability of finite-element discretizations of convection-diffusion-reaction equations*, IMA J. Numer. Anal. **31** (2011), no. 1, 147–164, DOI 10.1093/imanum/drp020. MR2755940

[Len00] W. Lenferink, *Pointwise convergence of approximations to a convection-diffusion equation on a Shishkin mesh*, Appl. Numer. Math. **32** (2000), no. 1, 69–86, DOI 10.1016/S0168-9274(99)00009-4. MR1724410

[Lin91] Q. Lin, *A rectangle test for finite element analysis*, Proc. Syst. Sci. Eng., Great Wall (H.K.) Culture Publish Co., 1991, pp. 213–216.

[Lin00] T. Linß, *Uniform superconvergence of a Galerkin finite element method on Shishkin-type meshes*, Numer. Methods Partial Differential Equations **16** (2000), no. 5, 426–440, DOI 10.1002/1098-2426(200009)16:5⟨426::AID-NUM2⟩3.3.CO;2-I. MR1778398

[Lin01] T. Linß, *The necessity of Shishkin decompositions*, Appl. Math. Lett. **14** (2001), no. 7, 891–896, DOI 10.1016/S0893-9659(01)00061-1. MR1849244

[Lin03] T. Linß, *Layer-adapted meshes for convection-diffusion problems*, Comput. Methods Appl. Mech. Engrg. **192** (2003), no. 9-10, 1061–1105, DOI 10.1016/S0045-7825(02)00630-8. MR1960975

[Lin04] T. Linß, *Error expansion for a first-order upwind difference scheme applied to a model convection-diffusion problem*, IMA J. Numer. Anal. **24** (2004), no. 2, 239–253, DOI 10.1093/imanum/24.2.239. MR2046176

[Lin10] T. Linß, *Layer-adapted meshes for reaction-convection-diffusion problems*, Lecture Notes in Mathematics, vol. 1985, Springer-Verlag, Berlin, 2010. MR2583792

[LK10] T. Linß and N. Kopteva, *A posteriori error estimation for a defect-correction method applied to convection-diffusion problems*, Int. J. Numer. Anal. Model. **7** (2010), no. 4, 718–733. MR2644301

[LMSZ09] F. Liu, N. Madden, M. Stynes, and A. Zhou, *A two-scale sparse grid method for a singularly perturbed reaction-diffusion problem in two dimensions*, IMA J. Numer. Anal. **29** (2009), no. 4, 986–1007, DOI 10.1093/imanum/drn048. MR2557053

[Lor81] J. Lorenz, *Zur Theorie und Numerik von Differenzenverfahren für Singuläre Störungen*, Ph.D. thesis, Universität Konstanz, 1981.

[LR06] G. Lube and G. Rapin, *Residual-based stabilized higher-order FEM for advection-dominated problems*, Comput. Methods Appl. Mech. Engrg. **195** (2006), no. 33-36, 4124–4138, DOI 10.1016/j.cma.2005.07.017. MR2229836

[LS99] T. Linß and M. Stynes, *A hybrid difference scheme on a Shishkin mesh for linear convection-diffusion problems*, Appl. Numer. Math. **31** (1999), no. 3, 255–270, DOI 10.1016/S0168-9274(98)00136-6. MR1711169

[LS01a] T. Linß and M. Stynes, *Asymptotic analysis and Shishkin-type decomposition for an elliptic convection-diffusion problem*, J. Math. Anal. Appl. **261** (2001), no. 2, 604–632, DOI 10.1006/jmaa.2001.7550. MR1853059

[LS01b] T. Linß and M. Stynes, *Numerical methods on Shishkin meshes for linear convection-diffusion problems*, Comput. Methods Appl. Mech. Engrg. **190** (2001), no. 28, 3527–3542. MR3618535

[LS12] R. Lin and M. Stynes, *A balanced finite element method for singularly perturbed reaction-diffusion problems*, SIAM J. Numer. Anal. **50** (2012), no. 5, 2729–2743, DOI 10.1137/110837784. MR3022240

[LSZ] Xiaowei Liu, Martin Stynes, and Jin Zhang, *Supercloseness of edge stabilization on Shishkin rectangular meshes for convection–diffusion problems with exponential layers*, IMA J. Numer. Anal., To appear.

[LU68] O. A. Ladyzhenskaya and N. N. Ural'tseva, *Linear and quasilinear elliptic equations*, Translated from the Russian by Scripta Technica, Inc. Translation editor: Leon Ehrenpreis, Academic Press, New York-London, 1968. MR0244627

[Mel02] J. M. Melenk, *hp-finite element methods for singular perturbations*, Lecture Notes in Mathematics, vol. 1796, Springer-Verlag, Berlin, 2002. MR1939620

[Mor96] K. W. Morton, *Numerical solution of convection-diffusion problems*, Applied Mathematics and Mathematical Computation, vol. 12, Chapman & Hall, London, 1996. MR1445295

[MOS12] J. J. H. Miller, E. O'Riordan, and G. I. Shishkin, *Fitted numerical methods for singular perturbation problems: Error estimates in the maximum norm for linear problems in one and two dimensions*, Revised edition, World Scientific Publishing Co. Pte. Ltd., Hackensack, NJ, 2012. MR2978532

[MS96] N. Madden and M. Stynes, *Linear enhancements of the streamline diffusion method for convection-diffusion problems*, Comput. Math. Appl. **32** (1996), no. 10, 29–42, DOI 10.1016/S0898-1221(96)00184-8. MR1426204

[MS97] N. Madden and M. Stynes, *Efficient generation of oriented meshes for solving convection-diffusion problems*, Int. J. Numer. Methods Engrg. **40** (1997), 565–576.

[MS12] N. Madden and M. Stynes, *A curious property of oscillatory FEM solutions of one-dimensional convection-diffusion problems*, Applications of mathematics 2012, Acad. Sci. Czech Repub. Inst. Math., Prague, 2012, pp. 188–196. MR3204411

[NSV18] T. A. Nhan, M. Stynes, and R. Vulanović, *Optimal uniform-convergence results for convection-diffusion problems in one dimension using preconditioning*, J. Comput. Appl. Math. **338** (2018), 227–238, DOI 10.1016/j.cam.2018.02.012. MR3773707

[OBB98] J. T. Oden, I. Babuška, and C. E. Baumann, *A discontinuous hp finite element method for diffusion problems*, J. Comput. Phys. **146** (1998), no. 2, 491–519, DOI 10.1006/jcph.1998.6032. MR1654911

[O'M14] R. E. O'Malley, *Historical developments in singular perturbations*, Springer, Cham, 2014. MR3289190

[OS86] E. O'Riordan and M. Stynes, *A uniformly accurate finite-element method for a singularly perturbed one-dimensional reaction-diffusion problem*, Math. Comp. **47** (1986), no. 176, 555–570, DOI 10.2307/2008172. MR856702

[OS91] E. O'Riordan and M. Stynes, *A globally uniformly convergent finite element method for a singularly perturbed elliptic problem in two dimensions*, Math. Comp. **57** (1991), no. 195, 47–62, DOI 10.2307/2938662. MR1079029

[OS08] E. O'Riordan and G. I. Shishkin, *Parameter uniform numerical methods for singularly perturbed elliptic problems with parabolic boundary layers*, Appl. Numer. Math. **58** (2008), no. 12, 1761–1772, DOI 10.1016/j.apnum.2007.11.003. MR2464809

[PW84] M. H. Protter and H. F. Weinberger, *Maximum principles in differential equations*, Springer-Verlag, New York, 1984. Corrected reprint of the 1967 original. MR762825

[QV94] A. Quarteroni and A. Valli, *Numerical approximation of partial differential equations*, Springer Series in Computational Mathematics, vol. 23, Springer-Verlag, Berlin, 1994. MR1299729

[Roo94] H.-G. Roos, *Ten ways to generate the Il'in and related schemes*, J. Comput. Appl. Math. **53** (1994), no. 1, 43–59, DOI 10.1016/0377-0427(92)00124-R. MR1305967

[Roo96] H.-G. Roos, *A note on the conditioning of upwind schemes on Shishkin meshes*, IMA J. Numer. Anal. **16** (1996), no. 4, 529–538, DOI 10.1093/imanum/16.4.529. MR1414845

[Roo12] H.-G. Roos, *Robust numerical methods for singularly perturbed differential equations: a survey covering 2008–2012*, ISRN Appl. Math. (2012), Art. ID 379547, 30. MR2999859

[RS15a] H.-G. Roos and M. Schopf, *An optimal a priori error estimate in the maximum norm for the Il'in scheme in 2D*, BIT **55** (2015), no. 4, 1169–1186, DOI 10.1007/s10543-014-0536-7. MR3434035

[RS15b] H.-G. Roos and M. Schopf, *Convergence and stability in balanced norms of finite element methods on Shishkin meshes for reaction-diffusion problems*, ZAMM Z. Angew. Math. Mech. **95** (2015), no. 6, 551–565, DOI 10.1002/zamm.201300226. MR3358551

[RS15c] H.-G. Roos and M. Stynes, *Some open questions in the numerical analysis of singularly perturbed differential equations*, Comput. Methods Appl. Math. **15** (2015), no. 4, 531–550, DOI 10.1515/cmam-2015-0011. MR3403449

[RST96] H.-G. Roos, M. Stynes, and L. Tobiska, *Numerical methods for singularly perturbed differential equations: Convection-diffusion and flow problems*, Springer Series in Computational Mathematics, vol. 24, Springer-Verlag, Berlin, 1996. MR1477665

[RST08] H.-G. Roos, M. Stynes, and L. Tobiska, *Robust numerical methods for singularly perturbed differential equations: Convection-diffusion-reaction and flow problems*, 2nd ed., Springer Series in Computational Mathematics, vol. 24, Springer-Verlag, Berlin, 2008. MR2454024

[RWG01] B. Rivière, M. F. Wheeler, and V. Girault, *A priori error estimates for finite element methods based on discontinuous approximation spaces for elliptic problems*, SIAM J. Numer. Anal. **39** (2001), no. 3, 902–931, DOI 10.1137/S003614290037174X. MR1860450

[RZ03] H.-G. Roos and H. Zarin, *The discontinuous Galerkin finite element method for singularly perturbed problems*, Challenges in scientific computing—CISC 2002, Lect. Notes Comput. Sci. Eng., vol. 35, Springer, Berlin, 2003, pp. 246–267, DOI 10.1007/978-3-642-19014-8_12. MR2070794

[San01] G. Sangalli, *A robust a posteriori estimator for the residual-free bubbles method applied to advection-diffusion problems*, Numer. Math. **89** (2001), no. 2, 379–399, DOI 10.1007/PL00005471. MR1855830

[San03] G. Sangalli, *Quasi optimality of the SUPG method for the one-dimensional advection-diffusion problem*, SIAM J. Numer. Anal. **41** (2003), no. 4, 1528–1542, DOI 10.1137/S0036142902411690. MR2034892

[Sch86] Friedhelm Schieweck, *Eine asymptotische angepaßte finite-element-methode für singulär gestörte elliptische randwertaufgaben*, Ph.D. thesis, Technische Hochschule Magdeburg, G. D. R., 1986.

[Sch98] Ch. Schwab, *p- and hp-finite element methods*, Numerical Mathematics and Scientific Computation, The Clarendon Press, Oxford University Press, New York, 1998. Theory and applications in solid and fluid mechanics. MR1695813

[Sch08] F. Schieweck, *On the role of boundary conditions for CIP stabilization of higher order finite elements*, Electron. Trans. Numer. Anal. **32** (2008), 1–16. MR2537213

[SE00] Y.-T. Shih and H. C. Elman, *Iterative methods for stabilized discrete convection-diffusion problems*, IMA J. Numer. Anal. **20** (2000), no. 3, 333–358, DOI 10.1093/imanum/20.3.333. MR1773263

[SGG99] R. Sacco, E. Gatti, and L. Gotusso, *A nonconforming exponentially fitted finite element method for two-dimensional drift-diffusion models in semiconductors*, Numer. Methods Partial Differential Equations **15** (1999), no. 2, 133–150, DOI 10.1002/(SICI)1098-2426(199903)15:2⟨133::AID-NUM1⟩3.3.CO;2-E. MR1674361

[Shi89] G. I. Shishkin, *Approximation of solutions of singularly perturbed boundary value problems with a parabolic boundary layer* (Russian), Zh. Vychisl. Mat. i Mat. Fiz. **29** (1989), no. 7, 963–977, 1102, DOI 10.1016/0041-5553(89)90109-2; English transl., U.S.S.R. Comput. Math. and Math. Phys. **29** (1989), no. 4, 1–10 (1991). MR1011021

[Smi85] D. R. Smith, *Singular-perturbation theory*, Cambridge University Press, Cambridge, 1985. An introduction with applications. MR812466

[SO97] M. Stynes and E. O'Riordan, *A uniformly convergent Galerkin method on a Shishkin mesh for a convection-diffusion problem*, J. Math. Anal. Appl. **214** (1997), no. 1, 36–54, DOI 10.1006/jmaa.1997.5581. MR1645503

[SRP13] K. K. Sharma, P. Rai, and K. C. Patidar, *A review on singularly perturbed differential equations with turning points and interior layers*, Appl. Math. Comput. **219** (2013), no. 22, 10575–10609, DOI 10.1016/j.amc.2013.04.049. MR3064568

[SS98] R. Sacco and M. Stynes, *Finite element methods for convection-diffusion problems using exponential splines on triangles*, Comput. Math. Appl. **35** (1998), no. 3, 35–45, DOI 10.1016/S0898-1221(97)00277-0. MR1605547

[SS09] G. I. Shishkin and L. P. Shishkina, *Difference methods for singular perturbation problems*, Chapman & Hall/CRC Monographs and Surveys in Pure and Applied Mathematics, vol. 140, CRC Press, Boca Raton, FL, 2009. MR2454526

[ST98] M. Stynes and L. Tobiska, *A finite difference analysis of a streamline diffusion method on a Shishkin mesh*, Numer. Algorithms **18** (1998), no. 3-4, 337–360, DOI 10.1023/A:1019185802623. MR1669942

[ST03] M. Stynes and L. Tobiska, *The SDFEM for a convection-diffusion problem with a boundary layer: optimal error analysis and enhancement of accuracy*, SIAM J. Numer. Anal. **41** (2003), no. 5, 1620–1642, DOI 10.1137/S0036142902404728. MR2035000

[Sty03a] M. Stynes, *A jejune heuristic mesh theorem*, Comput. Methods Appl. Math. **3** (2003), no. 3, 488–492, DOI 10.2478/cmam-2003-0031. MR2058042

[Sty03b] M. Stynes, *Numerical methods for convection-diffusion problems or the 30 years war*, 20th Biennial Conf. on Numerical Analysis (D.F. Griffiths and G.A. Watson, eds.), Numerical Analysis Report NA/217, University of Dundee, U.K., arXiv 1306.5172, pp. 95–103 (2003).

[Sty05] M. Stynes, *Steady-state convection-diffusion problems*, Acta Numer. **14** (2005), 445–508, DOI 10.1017/S0962492904000261. MR2170509

[VeBK95] A. B. Vasil′eva, V. F. Butuzov, and L. V. Kalachev, *The boundary function method for singular perturbation problems*, SIAM Studies in Applied Mathematics, vol. 14, Society for Industrial and Applied Mathematics (SIAM), Philadelphia, PA, 1995. With a foreword by Robert E. O'Malley, Jr. MR1316892

[Ver05] R. Verfürth, *Robust a posteriori error estimates for stationary convection-diffusion equations*, SIAM J. Numer. Anal. **43** (2005), no. 4, 1766–1782, DOI 10.1137/040604261. MR2182149

[Zha03] Z. Zhang, *Finite element superconvergence on Shishkin mesh for 2-D convection-diffusion problems*, Math. Comp. **72** (2003), no. 243, 1147–1177, DOI 10.1090/S0025-5718-03-01486-8. MR1972731

[Zho97] G. Zhou, *How accurate is the streamline diffusion finite element method?*, Math. Comp. **66** (1997), no. 217, 31–44, DOI 10.1090/S0025-5718-97-00788-6. MR1370859

[ZLY16] J. Zhang, X. Liu, and M. Yang, *Optimal order L^2 error estimate of SDFEM on Shishkin triangular meshes for singularly perturbed convection-diffusion equations*, SIAM J. Numer. Anal. **54** (2016), no. 4, 2060–2080, DOI 10.1137/15M101035X. MR3519135

[ZR05] H. Zarin and H.-G. Roos, *Interior penalty discontinuous approximations of convection-diffusion problems with parabolic layers*, Numer. Math. **100** (2005), no. 4, 735–759, DOI 10.1007/s00211-005-0598-1. MR2194592

[ZS11] L. Zhu and D. Schötzau, *A robust a posteriori error estimate for hp-adaptive DG methods for convection-diffusion equations*, IMA J. Numer. Anal. **31** (2011), no. 3, 971–1005, DOI 10.1093/imanum/drp038. MR2832787

[ZXZ09] Z. Zhang, Z. Xie, and Z. Zhang, *Superconvergence of discontinuous Galerkin methods for convection-diffusion problems*, J. Sci. Comput. **41** (2009), no. 1, 70–93, DOI 10.1007/s10915-009-9288-y. MR2540105

Index

Selected Published Titles in This Series

196 **Martin Stynes and David Stynes,** Convection-Diffusion Problems, 2018
192 **Tai-Peng Tsai,** Lectures on Navier-Stokes Equations, 2018
191 **Theo Bühler and Dietmar A. Salamon,** Functional Analysis, 2018
190 **Xiang-dong Hou,** Lectures on Finite Fields, 2018
189 **I. Martin Isaacs,** Characters of Solvable Groups, 2018
188 **Steven Dale Cutkosky,** Introduction to Algebraic Geometry, 2018
187 **John Douglas Moore,** Introduction to Global Analysis, 2017
186 **Bjorn Poonen,** Rational Points on Varieties, 2017
185 **Douglas J. LaFountain and William W. Menasco,** Braid Foliations in Low-Dimensional Topology, 2017
184 **Harm Derksen and Jerzy Weyman,** An Introduction to Quiver Representations, 2017
183 **Timothy J. Ford,** Separable Algebras, 2017
182 **Guido Schneider and Hannes Uecker,** Nonlinear PDEs, 2017
181 **Giovanni Leoni,** A First Course in Sobolev Spaces, Second Edition, 2017
180 **Joseph J. Rotman,** Advanced Modern Algebra: Third Edition, Part 2, 2017
179 **Henri Cohen and Fredrik Strömberg,** Modular Forms, 2017
178 **Jeanne N. Clelland,** From Frenet to Cartan: The Method of Moving Frames, 2017
177 **Jacques Sauloy,** Differential Galois Theory through Riemann-Hilbert Correspondence, 2016
176 **Adam Clay and Dale Rolfsen,** Ordered Groups and Topology, 2016
175 **Thomas A. Ivey and Joseph M. Landsberg,** Cartan for Beginners: Differential Geometry via Moving Frames and Exterior Differential Systems, Second Edition, 2016
174 **Alexander Kirillov Jr.,** Quiver Representations and Quiver Varieties, 2016
173 **Lan Wen,** Differentiable Dynamical Systems, 2016
172 **Jinho Baik, Percy Deift, and Toufic Suidan,** Combinatorics and Random Matrix Theory, 2016
171 **Qing Han,** Nonlinear Elliptic Equations of the Second Order, 2016
170 **Donald Yau,** Colored Operads, 2016
169 **András Vasy,** Partial Differential Equations, 2015
168 **Michael Aizenman and Simone Warzel,** Random Operators, 2015
167 **John C. Neu,** Singular Perturbation in the Physical Sciences, 2015
166 **Alberto Torchinsky,** Problems in Real and Functional Analysis, 2015
165 **Joseph J. Rotman,** Advanced Modern Algebra: Third Edition, Part 1, 2015
164 **Terence Tao,** Expansion in Finite Simple Groups of Lie Type, 2015
163 **Gérald Tenenbaum,** Introduction to Analytic and Probabilistic Number Theory, Third Edition, 2015
162 **Firas Rassoul-Agha and Timo Seppäläinen,** A Course on Large Deviations with an Introduction to Gibbs Measures, 2015
161 **Diane Maclagan and Bernd Sturmfels,** Introduction to Tropical Geometry, 2015
160 **Marius Overholt,** A Course in Analytic Number Theory, 2014
159 **John R. Faulkner,** The Role of Nonassociative Algebra in Projective Geometry, 2014
158 **Fritz Colonius and Wolfgang Kliemann,** Dynamical Systems and Linear Algebra, 2014
157 **Gerald Teschl,** Mathematical Methods in Quantum Mechanics: With Applications to Schrödinger Operators, Second Edition, 2014
156 **Markus Haase,** Functional Analysis, 2014